5th International Conference on Ship and Offshore Technology (ICSOT 2017)

Development in Ships Design and Construction

Depok, Indonesia
4-15 November 2017

ISBN: 978-1-5108-8307-9

Printed from e-media with permission by:

Curran Associates, Inc.
57 Morehouse Lane
Red Hook, NY 12571

Some format issues inherent in the e-media version may also appear in this print version.

The Royal Institution of Naval Architects
8-9 Northumberland Street
London, WC2N 5DA
United Kingdom

Phone: 020 7235 4622
Fax: 020 7259 5912

publications@rina.org.uk

Additional copies of this publication are available from:

Curran Associates, Inc.
57 Morehouse Lane
Red Hook, NY 12571 USA
Phone: 845-758-0400
Fax: 845-758-2633
Email: curran@proceedings.com
Web: www.proceedings.com

UNIVERSITAS INDONESIA
Veritas, Probitas, Iustitia
EST. 1849

PROCEEDINGS OF 5th INTERNATIONAL CONFERENCE ON SHIP AND OFFSHORE TECHNOLOGY (ICSOT)

The Royal Institution of Naval Architects

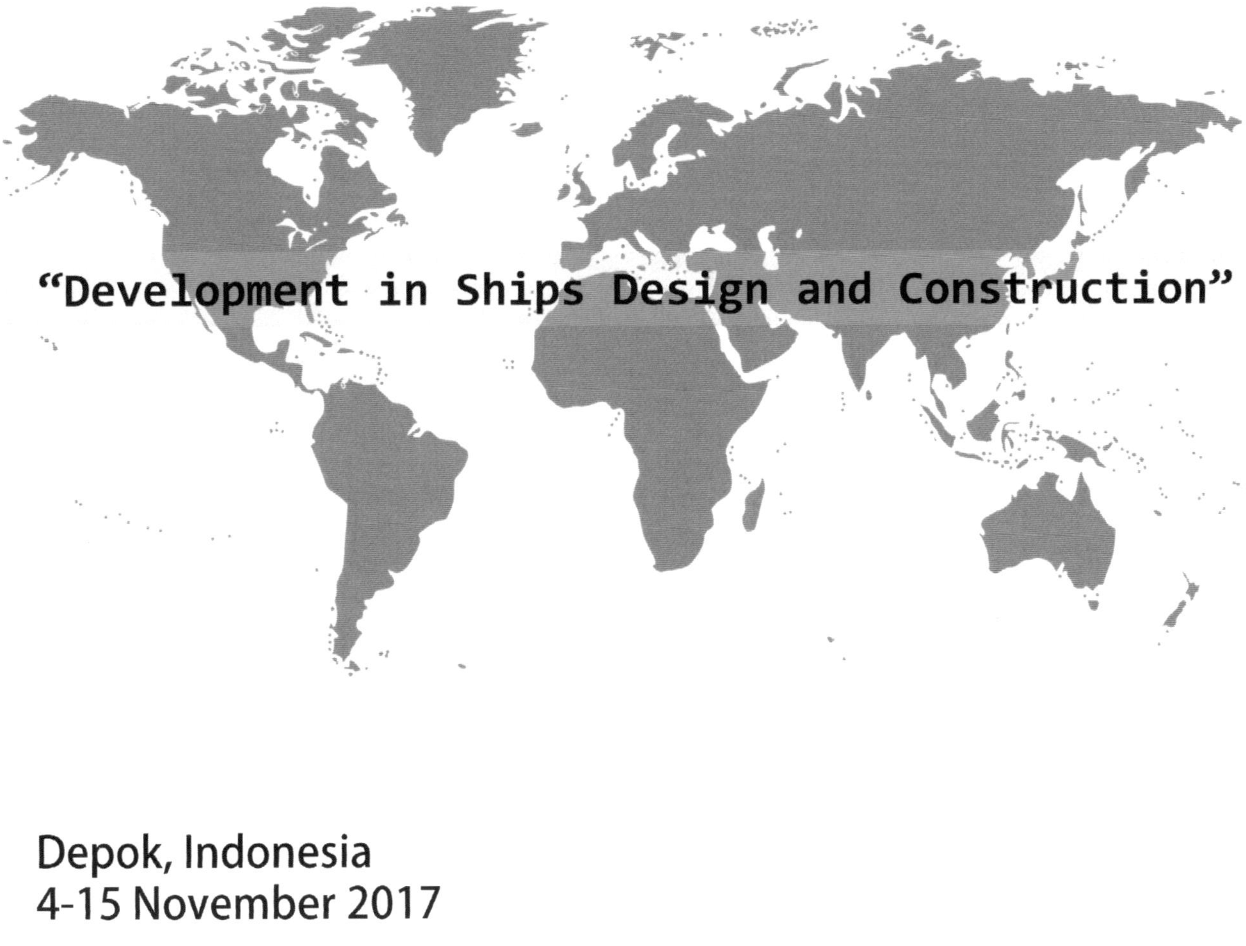

"Development in Ships Design and Construction"

Depok, Indonesia
4-15 November 2017

Organized by:

Program Studi Teknik Perkapalan Universitas Indonesia

RINA (Royal Institution of Naval Architects)

EDITORIAL BOARD

PREFACE

It gives me immense pleasure to know that Naval Architecture and Marine Engineering Study Programme, Mechanical Engineering Department has been organizing the 5th International Conference on Ship and Offshore Technology to be held at Faculty of Engineering Universitas Indonesia. This two days conference speaks to the importance of fostering networks among researchers, academics, and practitioners in the region and beyond to share knowledge and research findings in the field of ship and offshore technology.

Universitas Indonesia is the oldest and one of the top ranking universities in Indonesia. It was established during the Dutch colonial time in 1849, and the Faculty of Engineering in Jakarta campus (after the one in Bandung, and now become Institut Teknologi Bandung) started in 1964. Over the years the faculty has grown and contributes to global development of science and technology, including marine and maritime technology, and pacing itself to become part of the world class universities.

I welcome all the delegates and participants of the International Conference on Ship and Offshore Technology, and hope that you will enjoy your time and gain benefits from the conference. I wish ICSOT 2017 a major success.

Prof. Dr. Ir. Dedi Priadi DEA
Dean, Faculty of Engineering Universitas Indonesia

PREFACE

International Conference on Ship and Offshore Technology and more popularly known as ICSOT is a biannual conference and jointly organized by Indonesian universities and the Royal Institution of Naval Architects (RINA). The Conference was initiated at ITS in 2010 as part of 50 years anniversary of ITS and higher education in naval architecture in Indonesia, 100 years of PT DPS the oldest shipyard in Indonesia and 150 years of RINA. The second Conference was conducted at Pattimura University (Ambon) in 2012, the third Conference was held at Hasanuddin University (Makassar) in 2014, and the fourth Conference was held at Institut Teknologi Sepuluh Nopember (ITS) in 2015 as special conference in commemoration to the 10 years anniversary of RINA Indonesia. So far during those conferences, delegates and participants come from UK, Sweden, Norway, Holland, Belgium, Turkey, Australia, Malaysia, Singapore, Japan, and Indonesia. Now in 2017 Universitas Indonesia (UI) has been appointed to host the fifth Conference. With the fast development of science and technology the conference attempts to provide a platform for exchanging ideas, research results, and experiences in various areas of ships and offshore technology among researchers in the region and beyond.

The organizing Committee both local and international have worked very hard for about one year in preparing this conference. Generous involvement of Naval Architecture and Marine Engineering Study Programme, Mechanical Engineering Department, and Faculty of Engineering Universitas Indonesia that make this conference happened.

Appreciations also directed to all delegates and participants of this conference who have come from many places far and near, local and international, surely your contributions will bring benefits in expanding the science and technology.

I wish ICSOT 2017 a great success.

Dr. Ir. Sunaryo
Chairman ICSOT 2017

PREFACE

It is a pleasure to welcome representatives of the global maritime industry to the *International Conference on Ship & Offshore Technology, Indonesia 2017*, organised by the Royal Institution of Naval Architects.

I am sure you would agree that the maritime industry is facing many challenges at this time. I am sure you would also agree that key to meeting these challenges is innovation, whether in design, construction or operation.

The maritime industry is highly dependent on technology, and it is in technological development that innovation has had and will continue to have the greatest impact in providing ships and marine structures which cost less to design, build and operate, are safer, and are more sensitive to the environment.

The International Conference on Ship & Offshore Technology, Indonesia 2017 provides an opportunity for members of the maritime industry to be informed of and to discuss the technical developments which will lead to the innovation which is key to the industry's future success.

It is the role of the Royal Institution of Naval Architects as a leading international professional society to promote and facilitate the exchange of technical and scientific information on developments in the maritime industry. The Institution is proud of the reputation which it enjoys for organising successful conferences of the highest quality, addressing the latest developments in the maritime industry and attracting a wide range of informed speakers and delegates. I hope you will agree that the *International Conference on Ship & Offshore Technology, Indonesia 2017* shows this reputation to be deserved.

Trevor Blakeley
Chief Executive
Royal Institution of Naval Architects

Table of Contents

OPTIMIZATION OF PIPING SYSTEM ARRANGEMENT IN SERIES SHIPS USING THE MODULARIZATION CONCEPT

Gunawan, Kunihiro Hamada, Takuo Deguchi, Hiroshima University, Japan
Hiroyuki Yamamoto and Yasufumi Morita Tsuneishi Shipbuilding Co., Ltd., Japan

SUMMARY

Recently, the shipbuilding industry in the world market has become extremely the competitive. To compete in the global market, shipbuilders need to produce ships more efficiently. Therefore, the development of new methods like building series ships, design optimization, and modularization concepts is necessary. This paper presents a new approach for engine room design based on the modularization concept. The characteristics of the proposed method are as follows:

- Attention was paid to piping systems of multiple bulk carrier series of different sizes. The cost and weight of the piping system as well as the similarity and the commonness of the modules and arrangements were considered.
- The piping system design process was divided into two stages: module definition and arrangement design.
- To define an effective module that could be commonly used in different ships, a design structure matrix was adopted.
- In the arrangement design, an optimization system was developed using a genetic algorithm to obtain a similar pattern for module arrangement in multiple series ships with specific consideration toward the cost and similarity.

This study discusses the details of the above mentioned methods. In addition, the piping system of an actual ship was designed using the proposed method and its effectiveness was evaluated.

1. INTRODUCTION

Shipbuilders require improvements in production methods for cost reduction and reduced delivery time of the ships to remain in business. The various production concepts, like block division, modularization and building ships with a standard design are possible solutions for improve production. Engine room design, including the piping system, is a complex process; therefore, modularization of its design is an effective strategy to minimize the complexity of the system. In shipbuilding, the piping design inside the engine room, specifications of the piping system, and arrangement of the piping system differ significantly owing to differences in ship sizes. These differences occur because individual ship designs are developed exclusively owing to different equipment used by the ship owners as well as different engine room requirements.

The modularization of the engine room has been studied by a few researchers. Koga et al. used the design structure matrix (DSM) method to generate a modular division of the engine room of a ship [1] and evaluated the result using functional completeness and module independency. A few researchers used the modularization concept based on grouping the system components in an engine room [2-4]. Further, they proposed the concept of engine room modularization in a modular standard container, based on the function volume. Consequently, the engineer-experience-based module arrangement was created.

In the case of engine room arrangement, several researchers used algorithms and sequential coordination to optimize the machinery arrangement and pipe routing [5-8]. The focus of their research was on the optimization of the engine room arrangement, plant equipment layout, and pipe route. However, a modular design for engine room equipment was not part of these studies. In addition, all the research [1-8] focused on the modularization or arrangement of only one ship or several components; series ships were not considered in these studies.

However, in the automobile industry, the view on modularization is gradually changing. Nowadays, modularization tends to target the overall optimization for all car types. Consider the example of Nissan Motor's Common Module Family (CMF); it efficiently designs various models such as small cars, large cars, and SUV by changing the combination of the engine compartment, cockpit, front underbody, and rear underbody as a module unit. Therefore, modularization in the automobile industry is more extensive than the engine room design in the shipbuilding industry; hence, there is an opportunity to improve engine room design in shipbuilding by enlarging its boundary.

On the basis of the above discussions, this study proposes a new approach for engine room design based on the modularization of various types of series ships. Furthermore, the study proposes a module arrangement in the engine room optimized by the use of a genetic algorithm, considering the cost and similarity of arrangements as objective functions.

The characteristics of this study are as follows:

- The target ships are various types of series ships.
- The commonness of the modules for all ships is

considered.

- Similarity in the module arrangement for all ships is considered.
- The proposed concept is used in an actual ship design process and the effectiveness is evaluated.

Furthermore, the paper discusses the actual applications of the proposed method.

2. PROBLEM DEFINITION

2.1 TARGET SHIP

The details of the target ships that are a part of the research are as follows:

- Multiple series of bulk carrier with different capacities are considered. The respective sizes of each series are handymax (58000 DWT), panamax (82000 DWT) and over panamax (98000 DWT).
- A single series consists of a significantly large number of ships.
- The piping system in the engine room for each ship is classified into a fuel system, lubricating oil system, seawater system, freshwater system, compress air system, and steam system.
- Every ship has different components and arrangements in the piping system based on the requirements of the owner. Moreover, the sizes of the components depend on the sizes of the bulk carriers.

2.2 TARGET PROCESS OF THIS PAPER

The piping design process generally consists of following four stages (Figure 1):

2.2 (a) Owner Requirements

This refers the information about the owner's requirements or desires that need to be considered to fulfill the subject ship's mission. This information is associated with the existing ship's characteristics and operation. It consists of several aspects such as the principal characteristics, operational conditions, tank and piping system and cargo information.

2.2 (b) Piping Diagram

The piping diagram for each system is examined after designer collects the information from the owner. The piping diagram depicts the correlation of the equipment and the instrumentation. Figure 1 (b) shows the main components of the cooling freshwater piping system and flow information.

2.2 (c) Part Arrangements

This stage focuses on the spatial arrangement analysis of all parts. In general, part arrangement is usually performed based on a reference ship, designer experiences, and the theoretical optimal solution.

2.2 (d) Pipe Routing

Pipe routing design is important to eliminate crossover and interference in the pipe patch. The pipe patch cannot be determined before all the parts are arranged. Figure 1 (d) depicts the pipe routing among parts. The pipe routing design depends on the experiences of the designer and the number of parts.

Each ship has different part arrangements in accordance to the above stated flow design. This paper focuses on the part arrangements; therefore, the piping diagram is already fixed and pipe routing is not considered.

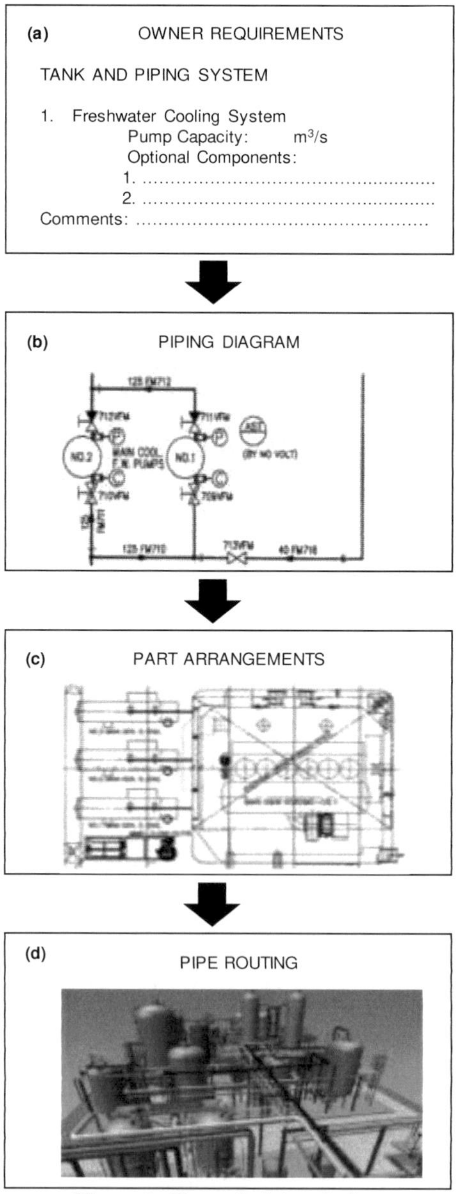

Figure 1: Flow of piping design

2.3 BASIC CONCEPT

The objective of this study is to develop the arrangement methodology of engine room parts by using modularization concept in consideration of the various kinds of series ships. To fulfil this objective, parts arrangement is divided into the following two problems.

2.3 (a) Module Definition Problem

Modularization involves grouping of parts with a strong dependency into one group. Modularization for common parts should be separate from that of the optional parts. The output of this process is a common module for various types of series ships and an optional parts module for a particular ship. The requirements for modularization are as follows:

- Module should be defined for a single ship, a series ships, and for various types of series ships.
- On the basis of the owner's requirements, both common and optional modules should be able to change in capacity and size without a change in the module configuration.
- A combination of common modules and optional modules should be possible to obtain a new ship type based on the owner's requirements.
- To use the modularization concept effectively, complex connections should be included in the module. Therefore, connections between the modules should be minimized.

2.3 (b) Module Arrangement Problem

Following the module definition process, the defined modules are arranged in the engine room. This process is known as the module arrangement problem. The following points are considered in the module arrangement problem:

- Various constraints such as the space requirement for maintenance, area for fixed components, etc. should be considered.
- Similarity of arrangements should be considered for a single ship, series ships, and for various types of series ships.
- Pipe costs should be minimized with respect to the pipe length, diameter, material, etc.

3. DETAILS OF MODULARIZATION

In this study, the module definition is executed by the following five steps. This section discusses the details of each step.

- Piping diagram (user input)
- E-R model for each ship
- E-R model for various types of series ships
- Modularization using DSM
- Modularization result

3.1 PIPING DIAGRAM (USER INPUT)

First, the piping diagram of all systems and for all series ships is required to define the module. In this study, the fuel system, lubricating oil system, seawater system, freshwater system, compress air system, and steam system are considered as the piping system.

3.2 E-R MODEL FOR EACH SHIP

To establish relationships between the various parts, this study uses the entity-relationship model (E-R model). The E-R model graphically represents the logical relationships of entities (objects). The model was first proposed by Peter Pin-Shan Chen of Massachusetts Institute of Technology in the 1970 [9]. In E-R modelling, the objects are represented by an entity, a relationship, and attributes; they can be defined as follows:

- An entity is a thing that exists either physically or logically. In this study, the entities are: the cooler, heater, purifier, filter, etc. The valve and branch are also considered entities.
- Relationships denote the manner in which the entities are related to one another. In this study, a relationship is the pipe between the entities.
- Attributes are the properties of entities. In this case, the flow capacity, heating value, part size, pipe diameter, etc. are considered attributes.

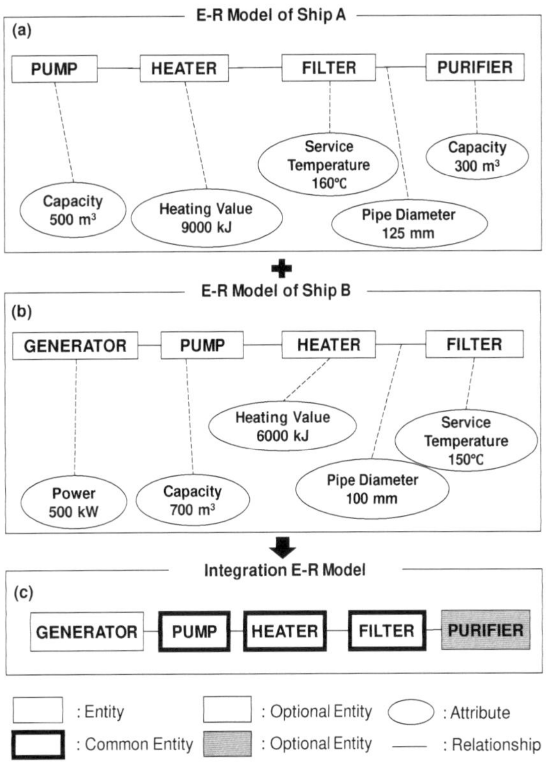

Figure 2: E-R model and its integration

Figures 2(a) & 2(b) depict the E-R model of a simple piping system. The entities are represented by the pump, heater filter, purifier, and generator. The following attributes of each entity are depicted as well: pump capacity (500 m^3), heater heating value (9000 kJ), filter service temperature (160°C), pipe diameter (125 mm), and purifier capacity (300 m^3). The relationship between the entities is represented by a pipe connection. The straight line in the figure is the pipe connection among the entities.

3.3 E-R MODEL FOR VARIOUS SERIES SHIPS

Subsequent to the generation of the E-R model for each ship, the E-R models are integrated into a single E-R model with specific focus on entities and their relationships. Figure 2 shows an example of such an integration. In Figure 2, although the attributes are different, the entities and relationship are the same. The entities, in this case, are estimated to be similar; therefore, similar entities and relationships are integrated into a single entity or single relationship. Thereafter, the entities and relationship are classified into the following two types.

- Common entities and relationship: these entities and relationship are used in all types of ships. In this case, the pump, hater, and filter are the common entities.
- Optional entities and relationships: these entities and relationship are used in a few of the ships, i.e. a series of ships, or ships for certain owners. In this case, the purifier is the optional entity in ship A and the generator is the optional entity in ship B.

Integration of E-R model is executed considering the requirements of each ship.

3.4 MODULARIZATION USING DSM

3.4 (a) Overview of DSM

DSM is a network modelling tool used to represent the elements in a system and their interactions. DSM is particularly well suited to applications for developing complex engineering systems and is currently being used primarily in the area of engineering management. The DSM is represented as a square N x N matrix and the interactions among the set of N elements are mapped. DSM has been effectively used to model several types of systems. In the product architecture, the DSM elements would be the components of the product and the interactions would be the interface between the components [10].

Figure 3 shows the flow of a DSM process. The E-R model consisting of 7 entities and 11 relationships is depicted in Figure 3(a). The relationships were set using weights. In this case, the two kinds of weight relationships are represented by the number 2 and 10. It should be noted that a stronger relationship is represented by a higher number. Once the weights were assigned, the weighted relationships were represented by a matrix (Figure 3(b)). The entities were represented by the vertical and horizontal matrix list, and the relationships were represented by the shell between the horizontal and vertical of the matrix list. The number weight was inserted to the appropriate shell. A clustering algorithm was used to create clusters by re-ordering the entity list to obtain a filled shell near the diagonal. In Figure 3(c), the clusters (modules) are shown in the black box. The larger number is collected inside the module and the smaller number is placed outside the cluster. The clustering results define the modules, as shown in Figure 3(d).

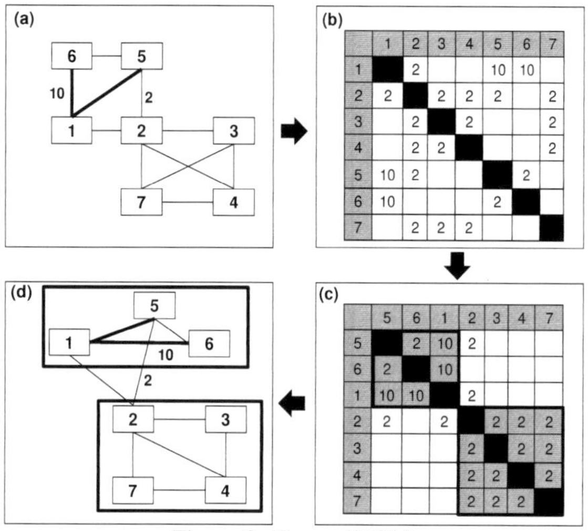

Figure 3: Flow of DSM

3.4 (b) DSM Application in This Study

In this study, clustering is performed using the weighted DSM. First, the matrices for common parts and optional parts are generated separately. Subsequently, the weights of the connections are set by the following rules:

- Connections for the common parts: The weight is assumed as the cost of the unit length of the corresponding pipe. When two or more pipes exist between the target parts, weight is assumed as the sum of the cost of corresponding pipes.
- Connections for optional parts: First, the weight is calculated in a similar manner to the common connections. Subsequently, the installation probability of the connections is multiplied with the weights.
- Once the weights are set, separate clusters of common parts and optional parts are prepared, thereby generating clusters (modules).

The flow of the DSM procedure in this study is shown in Figure 4.

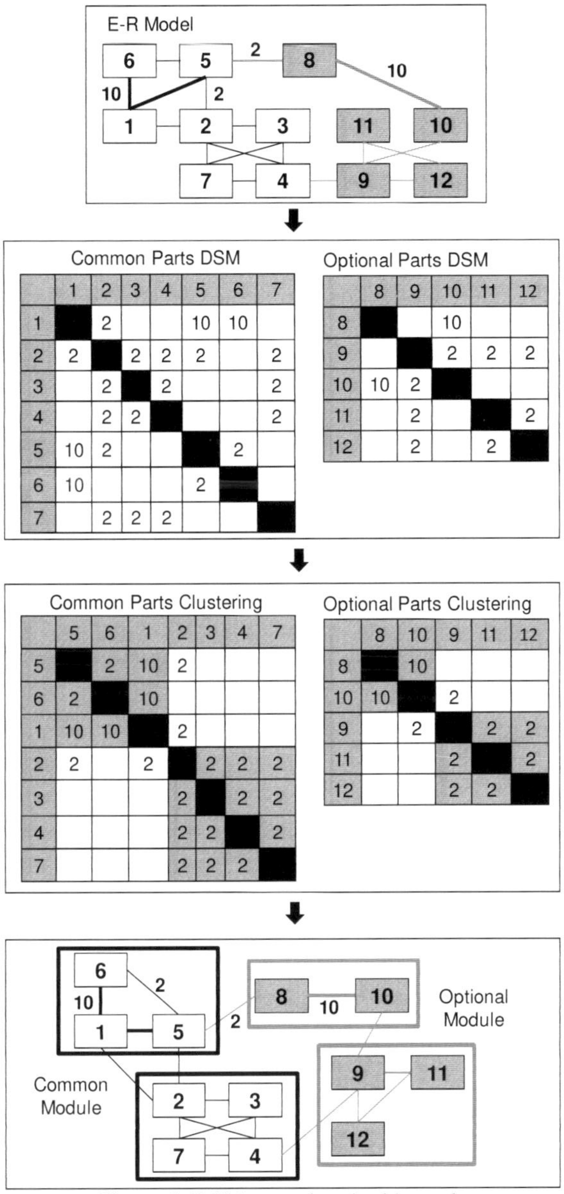

Figure 4: DSM procedure in this study

3.5 MODULARIZATION RESULT

In this study, each ship consists of six piping systems. Consider the example of seawater system; it comprises 233 common components and 8 optional components. Figure 5 shows the modularization result of the seawater system. The group components inside the black box are the common modules and the group components inside the coloured box depict the optional module. The red box denote the fixed parts, like the main engine, tank, and generator engine, and they should be situated outside the module.

In consonance with the module requirements, the seawater modularization can be established.

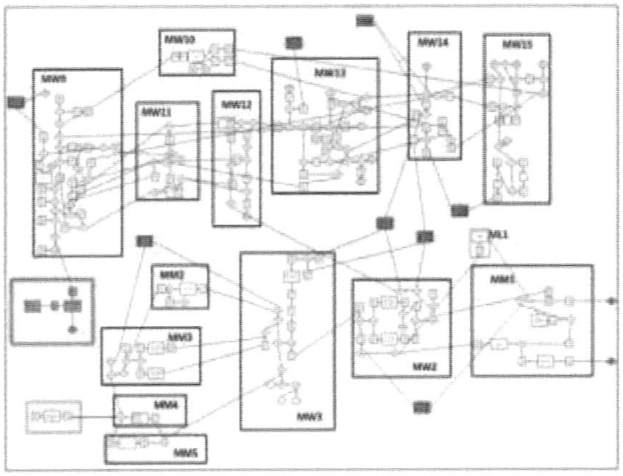

Figure 5: Modularization result of seawater system

The modularization was realized for a single ship or a series ships as well as for various types of series ships that can be established with 15 common modules and 2 optional modules. Each module can be changed in capacity or size without changes in the configuration. It is possible to configure a new piping system with a combination of all modules. In the above figure, there are 40 pipe connections among the modules and 233 pipe connections inside the module. Therefore, the proposed method satisfies all the requirements set in the section 2.3(a).

4. DETAILS OF ARRANGEMENT DESIGN

4.1 BASIC CONCEPT

In the arrangement design, modules defined in the previous section are arranged in the engine room. To realize the arrangement for series ships of different sizes, the following concepts were introduced in this study;

- Concurrent arrangements of three series ships were executed. Hence, various options could be considered at once.
- In the aforementioned case, the design space becomes relatively large; moreover, it is difficult to obtain an optimum solution within a limited time when the positions of modules are directly treated as design valuables. Therefore, decks in the engine room and the modules were divided into meshes.
- The cost of the pipes and similarity of arrangements were set as objective functions.
- A genetic algorithm was adopted for optimization.

First, the deck information and the module information were input by the user. Later, the genetic algorithm optimization technique was used on the data. The characteristics of the genetic algorithm, in this paper, are the chromosome representation and fitness calculation.

This section discusses the details of the above stated method.

4.2 INITIAL CONDITION (USER INPUT)

The deck information, module information, and the pipe information between modules were input by user before optimization. These were treated as the initial condition of the arrangement optimization.

4.2 (a) Deck Information

First, the deck information of all the target ships was input by the user in the design. Figure 6 shows the simplified example for deck information; the following information was input as deck information.

- The black area is the hull boundary of the ship.
- The gray area represents a few fixed parts such as the main engine, tank. The arrangement of the parts and pipe could not be accomplished.
- In the white area, all parts were arranged.

In this study, three series ships were set as targets with each ship having a three-deck structure. Therefore, information on nine decks was input as the initial condition. Consequently, the mesh size was set as 40 cm x 40 cm.

4.2 (b) Module Information

Module information comprises the information on the size and the limitations of the arrangement. In Table 1, the module A and B have a size of 6 columns × 4 rows and 5 columns × 5 rows, respectively. The modules A and B both should be arranged on the 3rd deck. Module A requires to be arranged on the portside and module B on the starboard side.

Table 1: Module information

ID	C	R	FLOOR	PARTIAL	3rd DECK	2nd DECK	SorP
A	6	4			1		S
B	5	5			1		P

Table 2: Pipe information

ID	A	B	C	D	E
A		1			
B	1		1		2
C		1		1	
D			1		
E		2			
ID	**A**	**B**	**C**	**D**	**E**
A		5.67			
B	5.67		3.24		2.45
C		3.24		8.12	
D			8.12		
E		2.45			

4.2 (c) Pipe Information

Pipe information comprises the number of pipe connections among modules and the weight per unit length data for each pipe. This data is used to calculate the pipe length and pipe weight. Table 2 gives an example of pipe information. As shown by the red circle in the table, there is only one pipe between module A and module B; the weight per unit length (kg/m) for this pipe is 5.67 kg/m.

4.3 DESIGN VARIABLES AND GENE REPRESENTATION

In this study, the basic unit of a chromosome is the arrangement information of one module for the three series ships. Figure 6 shows the gene sequence for a single module placement. The gene sequence is further divided into four parts; the first part is for the deck decision and the others are for module positions in the three series ships, i.e. 58BC, 82BC, 98BC, respectively. In Figure 6, the module position of 58BC is "7". The number is assigned to the free mesh on the decks in consideration of the fixed components and the pre-existing arranged modules (Figure 6). Therefore, the module is arranged at the mesh "7".

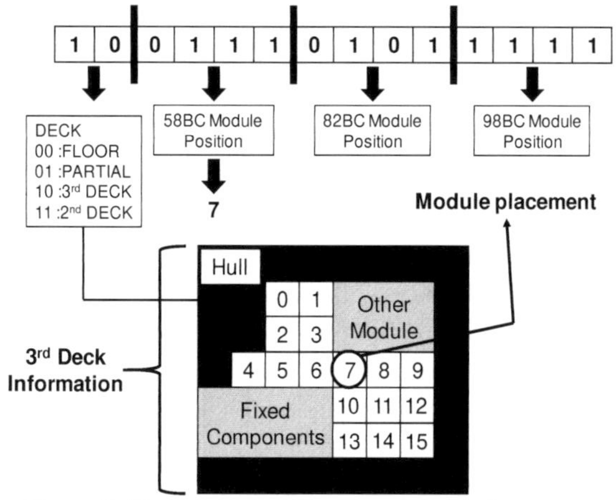

Figure 6: Gene representation for module placement

4.4 OBJECTIVE FUNCTION

The objective function in this study is the cost and the similarity of the arrangement for various types of series ships.

4.4 (a) Cost Calculation

First, the length of the pipe is calculated based on the following simple rules:

- The start and end of the pipes are considered as the centers of the connecting modules.
- In the case of the same deck, the number of meshes

between the start and end is counted. We count meshes for vertical and horizontal directions and not for diagonal directions.

- In the case of different decks, the height of the deck is added.

The cost of each pipe is calculated based on the length, material, and dimeter of the pipe. Subsequently, the costs of all pipes are summarized.

4.4 (b) Similarity Calculation

First, the position of the modules of each series is normalized by using the following equations (1).

$$(x, y) = \left(\frac{x_o}{sum\ of\ columns}, \frac{y_o}{sum\ of\ rows}\right) \quad (1)$$

x : *column position after normalize*
y : *row position after normalize*
x_o : *original column position*
y_o : *original row position*

Subsequently, similarity is calculated by using the following equation (2).

$$Similarity = \sum\left\{\sqrt{(x_1 - x_2)^2 + (y_1 - y_2)^2} + \sqrt{(x_2 - x_3)^2 + (y_2 - y_3)^2} + \sqrt{(x_3 - x_1)^2 + (y_3 - y_1)^2}\right\} \quad (2)$$

(x_1, y_1) : *module normalize coordinate of 58BC*
(x_2, y_2) : *module normalize coordinate of 82BC*
(x_3, y_3) : *module normalize coordinate of 98BC*

The equation denotes the summary of the relative distance of one module within the normalized deck area.

5. RESULT OF ARRANGEMENT

5.1 PROBLEM DEFINITION

To evaluate the effectiveness of this study, following optimizations were executed:

Case 1: Current arrangement in the company
Case 2: Optimization for one ship
Case 3: Optimization for one series of ships
Case 4: Optimization for various types of series ships of different sizes.

The target ships were set as the Bulk Carrier with capacities of 58000 DWT, 82000 DWT and 98000 DWT.

5.2 OPTIMIZATION RESULT

5.2 (a) Cost Comparison

The costs and pipe lengths of Cases 1–4 are compared in Table 3. The cost and pipe length of Case 1 were set as 100. As observed in the table, Case 2 is the most optimum in terms of the cost and pipe length. However, there are very slight differences between the values of Cases 3 and 4 in comparison to Case 2.

Table 3: Optimization result

Result (%)		Case 1	Case 2	Case 3	Case 4
58BC	Length	100	97.93	98.09	98.23
	Cost	100	98.27	98.29	98.42
82BC	Length	100	95.74	95.77	95.83
	Cost	100	96.32	96.33	96.38
98BC	Length	100	95.08	95.19	95.33
	Cost	100	96.49	96.53	96.62

5.2 (b) Comparison of Arrangements

Figure 7 shows the module positions of Cases 2–4. As shown in the figure, the module positions of the three series are different in Cases 2 and 3. However, in Case 4, the module positions of the three series are almost similar even though the sizes of the modules and the decks are different. A similar arrangement has benefit in terms of maintenance and the design/production lead time. Therefore, Tsuneishi Shipbuilding Co., Ltd, used the proposed arrangement for developing the ship.

6. CONCLUSION

In this study, a new piping system arrangement with respect to series ships and its modularization concept are proposed. First, the piping arrangement design was divided into two stages: module definition problem and module arrangement problem. To define an effective module that can be commonly used for different ships, the DSM method was adopted. Furthermore, for the arrangement design, an optimization system was developed using a genetic algorithm to obtain similar pattern of module arrangement for various types of series ships with specific consideration of the cost and similarity.

The following conclusions can be drawn:

- Modularization is an effective method to reduce the complexity of piping design in the engine room.
- Considering various types of series ships, a common modularization was obtained for common parts as well as for optional parts.
- To optimize the module arrangement design, a genetic algorithm was adopted.
- A similar arrangement design was obtained by using the proposed method for various types of series ships.

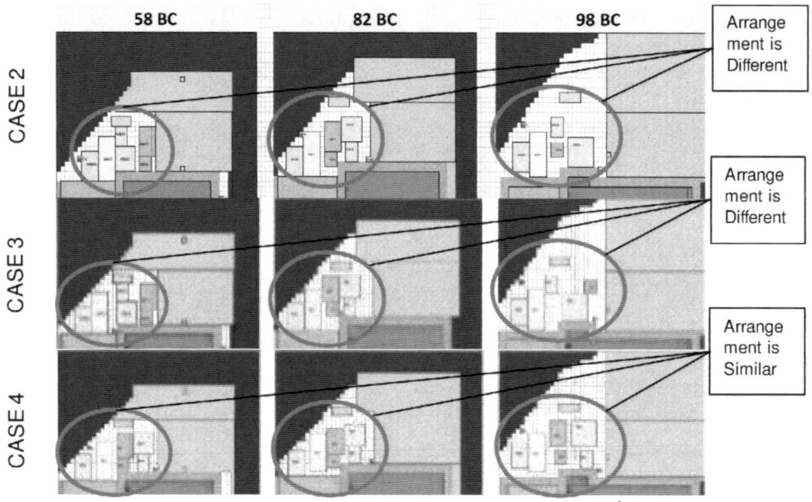

Figure 7: Differences of optimized arrangements in 3rd deck

7. REFERENCES

1. Tsuyoshi Koga, Takashi Niwa and Kazuhiro Aoyama, 'A Modular Design Method for an Engine Room Based on an Integrated Network Description and Division', 10th International Marine Design Conference, Trondheim, Norway 26 - 29 May 2009.

2. Ralf Baade, Friedrich Klinge, Kevin Lynaugh, Frank Woronkowicz and Klaus-Michael Seidler, 'Modular Outfitting', Journal of Ship Production, Vol. 14, No. 3, August 1998, pp. 170-179.

3. Peter E. Jaquith, Richard M. Burns, Steve E. Dunbarr, B. J. Fontaine, Harry A. Nelson and John L. Silveira, 'Modular Engine Room Design and Construction for the Strategic Sealift Ships', Journal of Ship Production, Vol. 12, No. 4, Nov. 1996, pp. 230-243.

4. Carlos Tomassoni, Ted Huynh and Thomas Schiller, 'Production-based Design Methodology for Shipboard Machinery Spaces', Journal of Ship Production, Vol. 19, No. 1, February 2003, pp. 53-63.

5. Wu, B.C., Young, G.S., Schmidt, W and Choppella, K, 'Applying Fuzzy Functions and Sequential Coordination to Optimization of Machinery Arrangement and Pipe Routing', Naval Engineers Journal, November 1998. pp. 43-54.

6. Dong-Myung Lee, Soo-Young Kim, Byung-Young Moon and Gyung-Ju Kang, 'Layout Design Optimization of Pipe System in Ship Engine Room for Space Efficiency', Journal of the Korean Society of Marine Engineering, Vol. 37, No. 7, 2013, pp. 784-791.

7. H. Kimura, 'Automatic Designing System for Piping and Instruments Arrangement Including Branches of Pipes', International Conference on Computer Applications in Shipbuilding 2011 (ICCAS 2011), Vol.3, pp.93-99.

8. Soo-Young Kim, Byung-Young Moon and Sung-Chul Shin, 'Evaluation Criterion of Machinery Arrangement Design in a Ship Engine Room', Journal of Ship Production, Vol. 25, No. 3, August 2009, pp. 117-125.

9. Chen, P. 'The Entity-Relationship Model toward an un-fined View of Data', ACM Transactions on Database Systems, Vol. 1, No. 1, March 1976, pp. 9-36.

10. Steven D. Eppinger and Tyson R. Browning, 'Design Structure Matrix Methods and Applications', MIT Press, 2012.

8. AUTHORS BIOGRAPHY

Gunawan is a Ph.D student at Hiroshima University under the guidance of Prof. Hamada.

Kunihiro Hamada holds the current position of professor at Hiroshima University, Japan. His major is design support system for shipbuilding.

Takuo Deguchi is a master course student at Hiroshima University under the guidance of Prof. Hamada.

Hiroyuki Yamamoto is Engineer at Tsuneishi Shipbuilding Co., Ltd, Japan.

Yasufumi Morita is Engineer at Tsuneishi Shipbuilding Co., Ltd, Japan.

A FRAMEWORK FOR THE PRELIMINARY DESIGN OF CONTAINER SHIPS CONSDIERING CARGO LOSS RISK IN SHIP-SHIP COLLISION ACCIDENTS

X Tan and **J Tao**, Maritime Institute at Nanyang Technological University, Singapore
D Konovessis, Singapore Institute of Technology, Singapore
Hock Eng Ang, School of Mechanical and Aerospace Engineering at Nanyang Technological University, Singapore

SUMMARY

This paper proposes a framework for the preliminary design of container ships considering the risk of cargo loss in ship-ship collision accidents. The procedure is an integration of a probabilistic collision damage model, common ship design software and parametric analytical models. Various ship design merits, e.g., the steel weight, ship resistance, cargo capacity, and the expected cargo loss due to ship-ship collisions are evaluated at the same time to achieve an optimal design in terms of the overall performance of the ship. The procedure was implemented in a parametric way, such that by systematically varying the parameters related to the conceptual design of the ship, large amount of design variants can be evaluated and compared. The comparison of the ship design merits across the different variants is discussed through a case study with a hypothetical container ship to get a preliminary insight into the trade-offs in the context of risk-based ship design.

NOMENCLATURE

List of Abbreviations:

IMO	International Maritime Organization
TEU	Twenty-Foot Equivalent Unit

List of Symbols:

B	Beam, moulded (container row)
BHW	Gap size between two cargo tanks (m)
C	Consequence due to an accident
C_B	Block coefficient
D	Depth, moulded (main deck) (container tier)
Disp.	Ship Displacement (ton)
HDB	Double bottom height (m)
HDS	Width of double skin sides (m)
L_{BP}	Length between perpendiculars (m)
L_{OA}	Length overall (m)
LCB	longitudinal center of buoyancy (m)
p	probability of having a certain amount of TEU loss
R	Risk metric (TEU)
T	Draft (m)
TEU_{damage}	Number of containers damaged due to water ingress

1. INTRODUCTION

Although the human and the hardware (in terms of ship capsizing for loss of stability) safeties of container ships in accidents has been extensively researched (*e.g.*, IMO's FSA report on container ships [1] and work by Wang and Foinikis [2]), The risk of cargo loss has seldom been touched upon. The consequence of container loss is not only relevant to the owner of the cargos, but also drifting containers can endanger other ships sailing in the same sea area. Thus, the design of container ships even from the initial preliminary phase should take into consideration of such issues and ensure a risk of cargo loss that is as low as reasonably possible.

Ship-ship collision is a complicated physical processes where uncertainties are present at every single stages of the interaction between the two ships [3] [4]. In order to quantitatively assess the collision risk and to obtain a statistically sound cargo loss distribution, extensive amount of collision cases have to be evaluated for each design variant. Since even the preliminary design optimization can easily involve hundreds of hull variants, methods such as performing integrated dynamic simulations with the finite element method and rigid body dynamics to collect the damage data to establish the damage distributions directly are considered unsuitable for the purpose of risk-based ship design and optimization. Although historical accident statistics are harvested from the real world, it reflects only fragments of reality. Relying solely on accident statistics may lead to biases either because they cannot reflect the future or because they do not take into consideration of specific conditions, e.g. ship structural design, for different cases [5] [6].

This paper introduces a scheme for the preliminary design and optimization of containerships considering the risk of cargo loss due to ship-ship collision accidents. The procedure is a synthesis of various first-principle-based analysis models and common naval architecture software. Risk is treated as a design objective as opposed to a constraint to achieve an optimal design that is a result of balancing it with other conventional ship design merits such as steel weight, calm water resistance and cargo capacity. The variables that define each design variant are the main particulars of the vessel and the inner tank arrangement parameters. In order to evaluate and compare safety level across the design variants, quantification of the risk of accidental in-hold cargo loss due to water ingress has become an important subject.

Risk is in the present work expressed as an expected cargo loss following [7] and [8]:

$$R = \sum_{i} p_i \cdot C_i \qquad (1)$$

where,
i is the index of consequences, which in the present paper represent the different numbers of cargo loss, represented by number of damaged TEUs due to water ingress;
C is the number of damaged TEUs due to water ingress in cargo tanks;
p is the probability that a certain number of TEU damage occurs.

The scope of the present work is listed as follows:

- Only in-hold cargo loss due to water ingress is considered, no falling overboard of on-deck containers is considered;
- All containers are considered damaged in a breached tank;

Since the focus of the present paper is on establishing a framework and generic methodology for the procedure of preliminary container ship design, the integrated analysis models can be considered as surrogate modules which can at a future stage be substituted when more accurate and refined models are available.

2. METHODOLOGY

2.1 OVERVIEW

The procedure consists of a number of analysis models, common naval architecture software and a probabilistic collision damage model. The idea is to establish a procedure that has high computational efficiency and at the same time can still take into consideration of details such as ship's main particulars and randomly distributed collision scenario variables. Parametric ship hull model was setup using NAPA. Conventional ship performance aspects such as the intact and damage stability, calm water resistance, cargo capacity and steel weight were calculated using NAPA too. In parallel, Monte Carlo simulations were conducted to obtain the probabilistic collision damage for each design variant. The main particulars and inner tank arrangement parameters were fed into the probabilistic damage model together with the randomly generated collision scenario variables. A schematic illustration of the procedure is given in Figure 1. In the following part of this section, the procedure and the adopted models are briefly introduced.

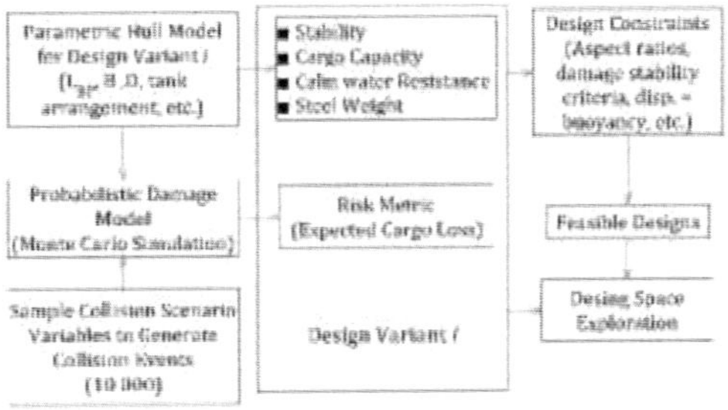

Figure 1: Schematic illustration of the procedure

2.2 RESISTANCE, STEEL WEIGHT AND CARGO CAPACITY

At the initial stage of ship design, empirical parametric models for resistance and steel weight are employed to calculate the global performance of the vessel. The weight component accounted for are the hull steel, outfit and machinery weights [9]. The resistance at the design speed was calculated using Holtrop and Mennen's formulae [10].

2.3 PROBABILISTIC COLLISION DAMAGE

The probabilistic ship-ship collision damage model given by [11] was adopted and implemented numerically in the present study to evaluate the probabilistic collision damage and subsequent in-hold cargo loss due to water ingress. The model is a regressional meta-model by processing simulation results of 80 000 collision cases for four different oil tanker designs. The simulations were conducted by using a deterministic numerical procedure of ship-ship collisions developed by [12]. 600 collision cases were generated based on each input collision scenario variable's probabilistic distribution. The same 600 collision cases were applied to each design variant and Monte Carlo simulations were conducted. By knowing the damage properties and inner tank arrangement, the breached tanks can then be identified and statistics of cargo loss due to water ingress can be calculated accordingly.

3. CASE STUDY

The established framework is applied to the preliminary design and optimization of a container ship considering the risk of cargo loss as well as other conventional ship design merits such as steel weight, TEU capacity and calm water resistance. Main particulars and parameters that control the cargo tank arrangement were varied systematically so that the design space was explored efficiently and effectively. For each feasible design variant, the expected in-hold cargo loss was calculated using the probabilistic collision damage model introduced in Section 2.3. Intact and damage stability, steel weight and calm water resistance were calculated in NAPA. Finally, the trade-offs among the design merits including risk were discussed.

3.1 REFERENCE SHIP

A hypothetical representative container ship with a nominal TEU capacity of 7500 is adopted in the present study as reference design. The main particulars of the ship are listed in Table 1.

Parameter	Value	Unit
L_{OA}	323.83	m
L_{BP}	293.90	m
B	42.85	m
D	24.71	m
C_B	0.65	/
LCB (to AP)	137.48	m
HDS	2.30	m
HDB	2.10	m
T	12.05	m
Disp.	101851	ton
Nominal TEU Capacity	7500	/

Table 1: Main particulars of the reference ship

3.2 DESIGN SPACE AND FEASIBLE DESIGNS

The bounding limits for the design variables which define the design space are summarized in Table 2.

Variable	Lower bound	Upper bound
L_{BP}	290.5 m	293.9 m
B	15 row	16 row
LCB	131.1 m	139.6 m
C_B	0.6	0.8
D	9 tier	10 tier
HDB	1.8 m	2.5 m
HDS	2.0 m	2.5 m
BHW	0.8 m	1.0 m

Table 2: Ranges of ship design variables that bound the design space

In Table 2, the beam and the depth of the vessels are represented by the number of containers in each corresponding direction. In the present study, the length of the ship in terms of container number in the longitudinal direction was not varied. The variation in L_{BP} as shown in Table 2 is due to the change in the gap sizes, BHW, between each two adjacent cargo tanks.

In order to explore the design space effectively and efficiently, low variance sequence was used to fill the design space by generating the design variables within the bounds given in Table 2.

The design constraints, with which feasible designs in the investigated design space can be identified, are listed below:

- The ratio B/D is limited to between 2.5 and 3.75;
- IMO Intact stability criteria [13];
- IMO damage stability code [14];

In the present study, originally 6 500 design variants were generated, of which 4 826 are feasible ones that fulfill the constraints listed above. The scatter plots of the generated ship design variables for the entire design space as well as for the feasible designs are given in Figure 2 appended at the back of the paper.

By comparing the two plots in Figure 2, it can be seen that, in general, the designs that have been filtered out have a large draft relative to the beam, which can lead to a less stable geometry under intact conditions. In addition, the draft of those designs are also relatively large compared with the depth, implying a smaller freeboard and hence higher probability of losing stability under damaged conditions due to water ingress through the hatches on deck.

3.3 PARAMETRIC CONTAINER SHIP MODEL

A parametric ship hull model (Figure 3) was established with NAPA. As the detailed information as to the lines of the ship is not available, a generic tanker geometry provided by NAPA was used. The design variables of the ship model were varied according to the data points as presented in Figure 2 a).

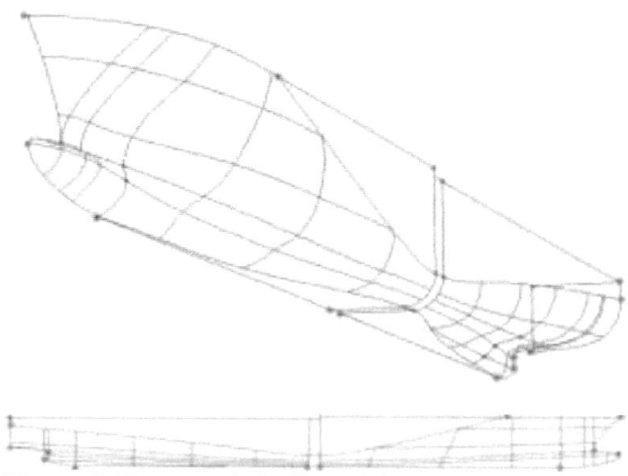

Figure 3: Parametric hull model in NAPA (the points that control the geometry of the hull are highlighted.)

3.4 LOADING CONDITION

In the present study, a homogenous weight of 10 tons for the containers was used. The vessel was first loaded up to the visibility line, then intact and damage stability criteria were checked. If the criteria were not fulfilled, then a tier of on-deck containers were removed, this process is repeated until the criteria were fulfilled. Ballast tanks are empty except for the need to adjust the vessel to an upright position.

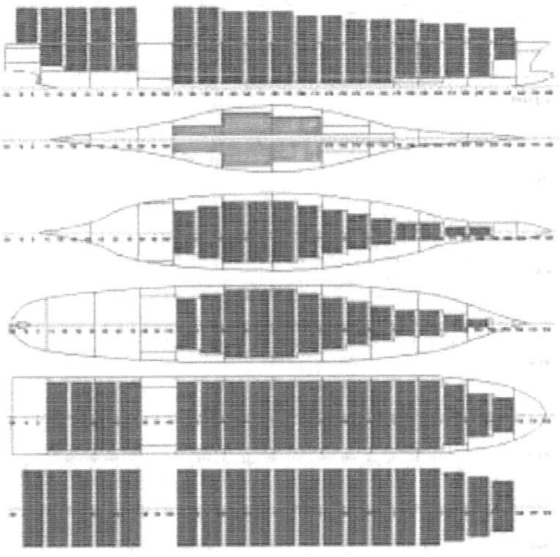

Figure 4: Side and plan views of the vessel with stacked containers (red blocks)

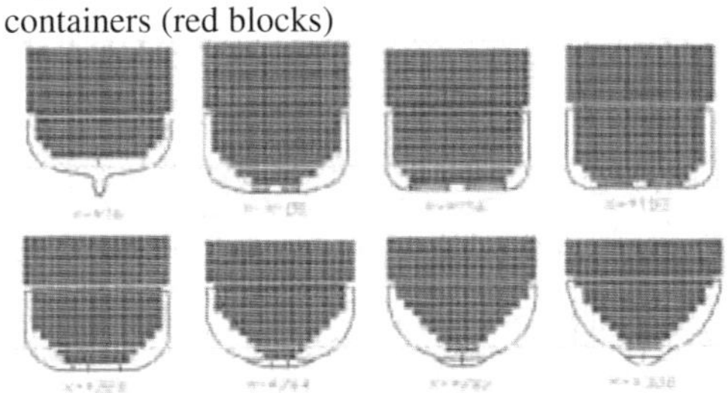

Figure 5: Front view of the vessel at different cross sections with stacked containers (red blocks)

3.5 RESULTS

In order to achieve an overall optimal design, multiple ship design merits were evaluated simultaneously, namely, the steel weight, TEU capacity, calm water resistance and the risk of cargo loss due to water ingress following a collision accident. This is discussed in this section over the simulated results.

Figure 6 (appended at the back of the paper) shows the scatter plots between each two of the simulated outputs, i.e. the design merits. The four different ship size categories due to the four possible combinations of B and D values are color coded. For each size category, two distinct magnitudes of on-deck container numbers can be identified. This is due to the limitation of the current strategy used in the model to remove on-deck containers to fulfill the stability criteria. When stability is not fulfilled, an entire tier is removed. This also lead to two distinct values of draft and thus resistance. The figure also shows that as the size of the ship in terms of container numbers grows, the steel weight grows, and so does the risk of cargo loss as described by the expected TEU damage. This trend is as expected.

Figures 7 through 9 shows the scatter plots among the design merits. The designs were ranked according to their utility values combined with different weights. It can be seen that the resistance and TEU capacity values are clustered around several discrete values. This is due to the fact that the sizes of the ship are clustered around four categories, two different beams and two different depths as given in Table 2, and the ship size for container ships is primarily a staircase function of the beam and depth.

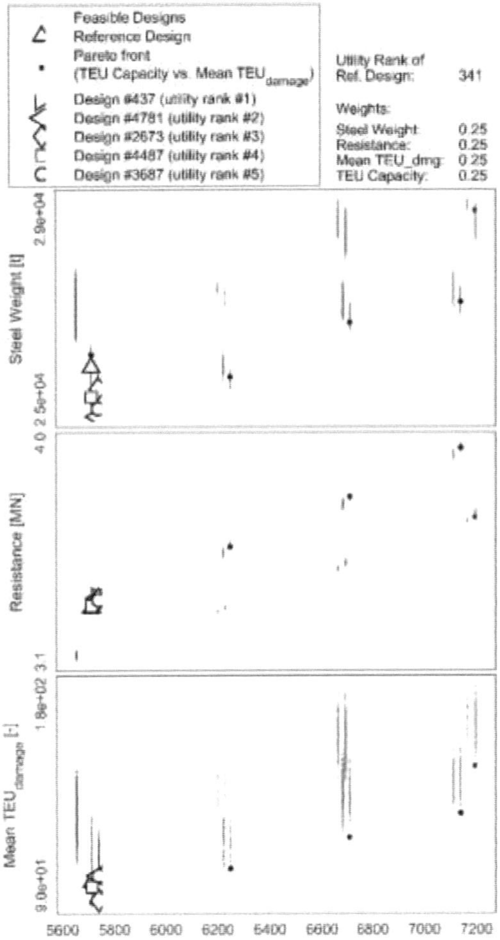

Figure 7: Scatter plots for the design merits (equal weights)

In Figure 7, equal weights are put on the four design merits. The five most highly ranked designs for this setting are concentrated in the smaller size segment, which is reasonable as all the merits but TEU capacity favor a smaller ship in general.

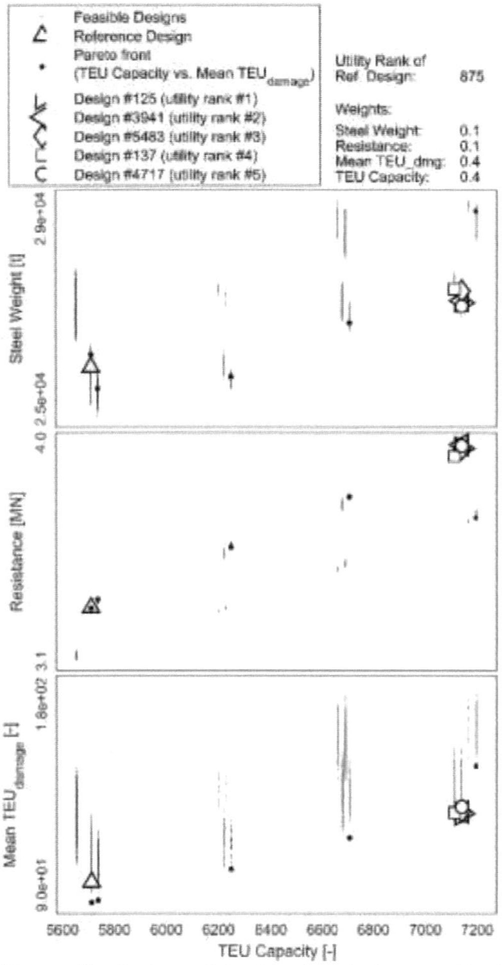

Figure 8: Scatter plots for the design merits (more weights on TEU damage risk and TEU capacity)

In Figure 8 and 9, more weights are put on TEU capacity and TEU damage risk. It is then seen that the top ranked designs now lie on the "Pareto front" between the TEU capacity and the TEU damage risk, which is the non-dominated set of designs. Also can be seen is that when preferences are given, larger designs can become competitive.

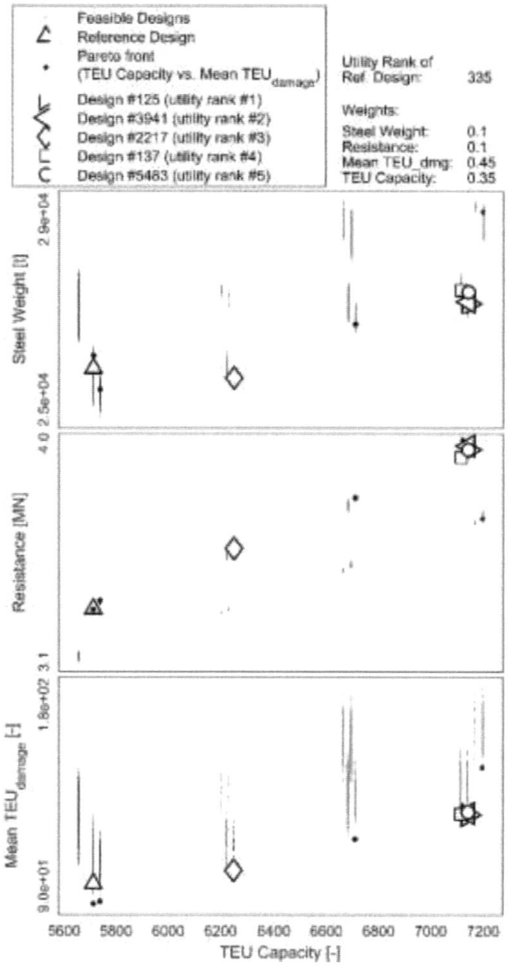

Figure 9: Scatter plots for the design merits (more weights on TEU damage risk)

4. CONCLUSIONS AND DISCUSSIONS

This paper proposes a computational framework for the preliminary design and optimization of container ships. The framework is an integration of various analytical models, a probabilistic model and common ship design software. The application of the framework was investigated through a case study with the preliminary design of a container ship with a 7000 ~ 8000 nominal TEU capacity. Four design merits including risk of cargo loss and the ones treated in conventional ship design process were evaluated simultaneously to achieve an overall optimal design.

It was shown that as the ship size grows, the steel weight, resistance, cargo loss risk as well as cargo capacity grows, as expected. Since only the increase of the cargo capacity is the desirable thing among the four design merits, trade-off needs to be made to arrive at the optimal design. The study showed that by using the utility function technique with different weight specification, different design variants can surface up as optimal. Also several designs have exhibited better overall performance as compared with the reference design. The framework can help facilitate decision making when a particular operational setting and design requirement is given.

In the present study, the number of the on-deck containers are restricted by the visibility line and stability criteria. When the stability criteria are not fulfilled, an entire tier is removed, until the criteria is fulfilled. Future work needs to be carried out to develop a strategy to remove on-deck containers individually.

5. ACKNOWLEDGEMENTS

This research is supported by Singapore Maritime Institute.

6. REFERENCES

1. IMO MARITIME SAFETY COMMITTEE, 'FSA – Container Vessels', MSC 83/INF.8, 2007.

2. WANG, J. and FOINIKIS, P., 'Formal Safety Assessment of Containerships', Marine Policy 25, 2001.

3. GOERLANDT, F., MONTEWKA, J., 'A Framework for Risk Analysis of Maritime Transportation Systems: A Case Study for Oil Spill from Tankers in A Ship-Ship Collision', Safety Science, Volume 76, Pages 42-66, 2015.

4. SORMUNEN, O. E., GOERLANDT, F., HÄKKINEN, J., et al., 'Uncertainty in Maritime Risk Analysis: Extended Case Study on Chemical Tanker Collisions', Proceedings of the Institution of Mechanical Engineers, Part M: Journal of Engineering for the Maritime Environment, Volume: 229 issue: 3, page(s): 303-320, 2014.

5. BROWN, A., TIKKA, K., DAIDOLA, J. C., LÜTZEN, M., and CHOE, I-H, 'Structural Design and Response in Collision and Grounding', In Annual meeting. Vancouver: SNAME, 2000.

6. DNV, 'Risk Assessment – Large Passenger Ships – Navigation', DNV Report 2003-0277, Annex II, 2003.

7. ISO, 'Risk Management Vocabulary, Guidelines for Use in Standards, ISO/IEC Guide. International Standards Organisation, Geneva. ISO Guideline vol 73, 2002.

8. VINNEM, J., 'Offshore Risk Assessment Principles, Modelling and Applications of QRA Studies', Springer-Verlag London, 2014.

9. WATSON, D.G.M., and GILFILLAN, A.W., 'Some Ship Design Methods', Transactions RINA, Vol. 119, 1977.

10. HOLTROP, J., and MENNEN, G.G.J., 'An Approximate Power Prediction Method', International Ship Building Progress, 1982.

11. VAN DE WIEL, G. and VAN DORP, J.R., 'An Oil Outflow Model for Tanker Collisions and Groundings', Annals of Operations Research, 187, 279–304, 2011.

12. BROWN, A. 'Alternative Tanker Designs, Collision Analysis', NRC Marine Board Committee on Evaluating Double-Hull Tanker Design Alternatives, 2001.

13. IMO, 'International Code on Intact Stability', RESOLUTION MSC.267(85), 2008.

14. IMO, 'International Convention for the Safety of Life at sea, Chapter II-1, Part B-1, Regulation 7', 1974.

7. AUTHORS BIOGRAPHY

Xiang Tan is Research Fellow at the Maritime Institute at Nanyang Technological University, Singapore. She graduated with MSc in Naval Architecture and Ocean Engineering from Tianjin University and PhD in Marine Technology from the Norwegian University of Science and Technology (NTNU). Her research interests include dynamics and the design of ships and marine structures under ice and wave actions. She is responsible for establishing the procedure, analyzing the results and writing the paper.

Jidong Tao is Research Associate at the Maritime Institute at Nanyang Technological University, Singapore. He is responsible for preparing the tanker model in NAPA and generating results of technical and economic performance of various tanker designs. His previous experience includes ship performance, structural design and analysis of ship and offshore engineering, etc.

Dimitrios Konovessis is Associate Professor at the Singapore Institute of Technology. He graduated with MEng in Naval Architecture and Marine Engineering from the National Technical University of Athens and PhD in Ship and Marine Technology from the University of Strathclyde. His research interests include design methods for ship safety, first-principle and performance-based approaches for risk-based ship design, operation and regulation, marine systems reliability and maintainability, and energy efficiency and management.

Hock Eng Ang graduated with a BSc in Mechanical Engineering from the University of Wales, Swansea and a MSc degree in Mechanical Engineering, with DIC, from University of London, Imperial College. He is Chartered Mechanical Engineer (UK) and Registered Professional Engineer with Practising Certificate (Malaysia). He is the academic focus coordinator for the Naval Architecture & Marine Engineering specialization courses offered by School of Mechanical & Aerospace Engineering, Nanyang Technological University, Singapore.

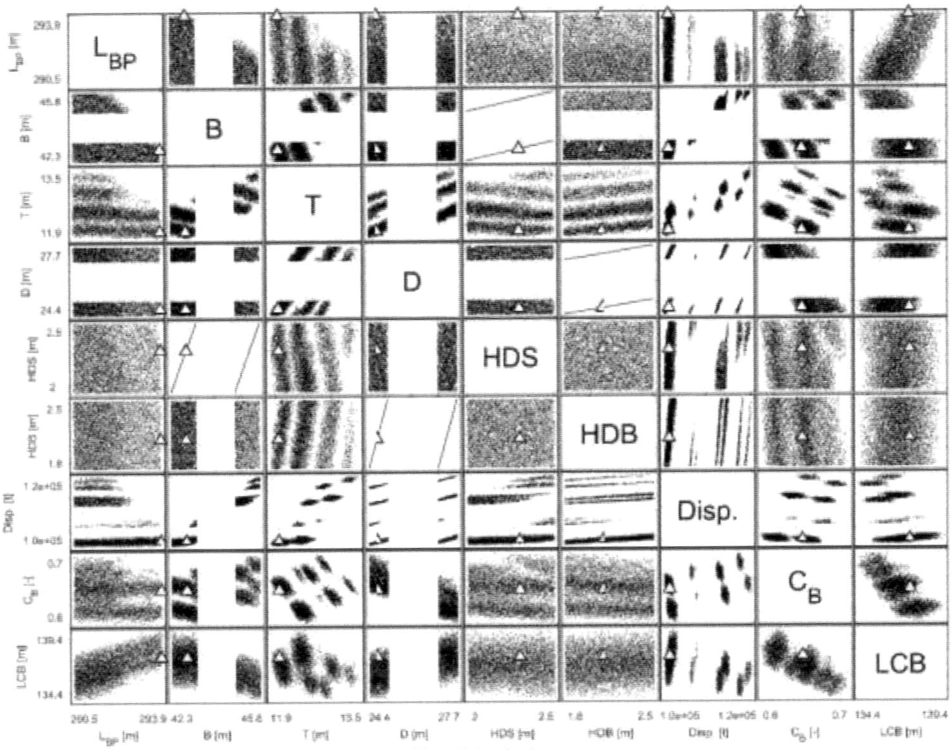

a) Feasible designs

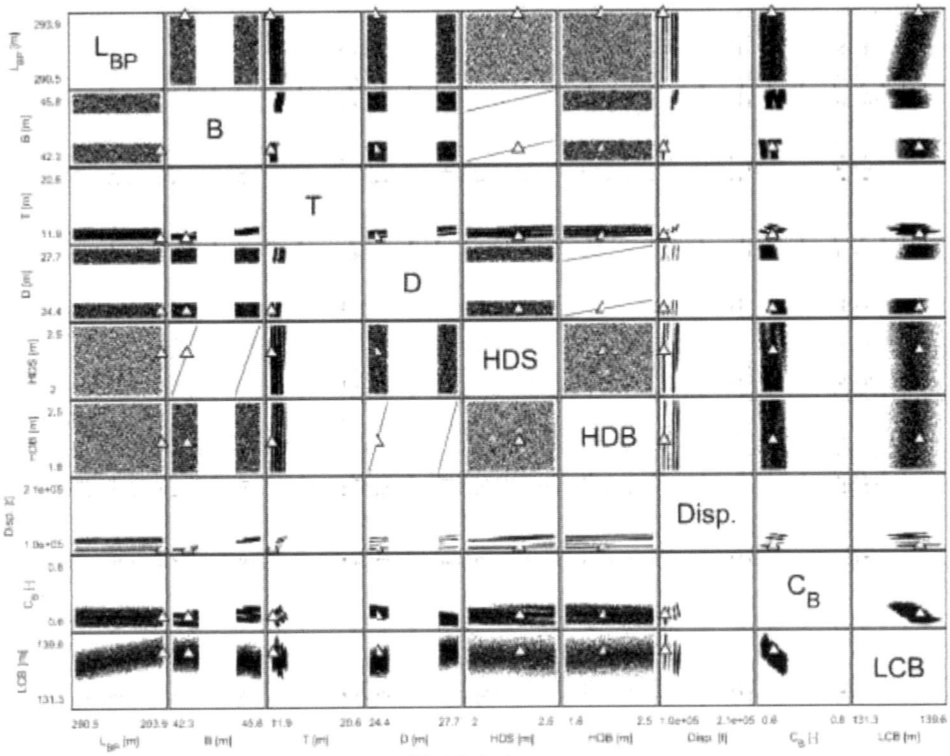

b) All designs

Figure 2: Input variable scatter plots (the points corresponding to the reference design are highlighted with white triangles; the upper and lower bounds of the data set are labeled along the axes)

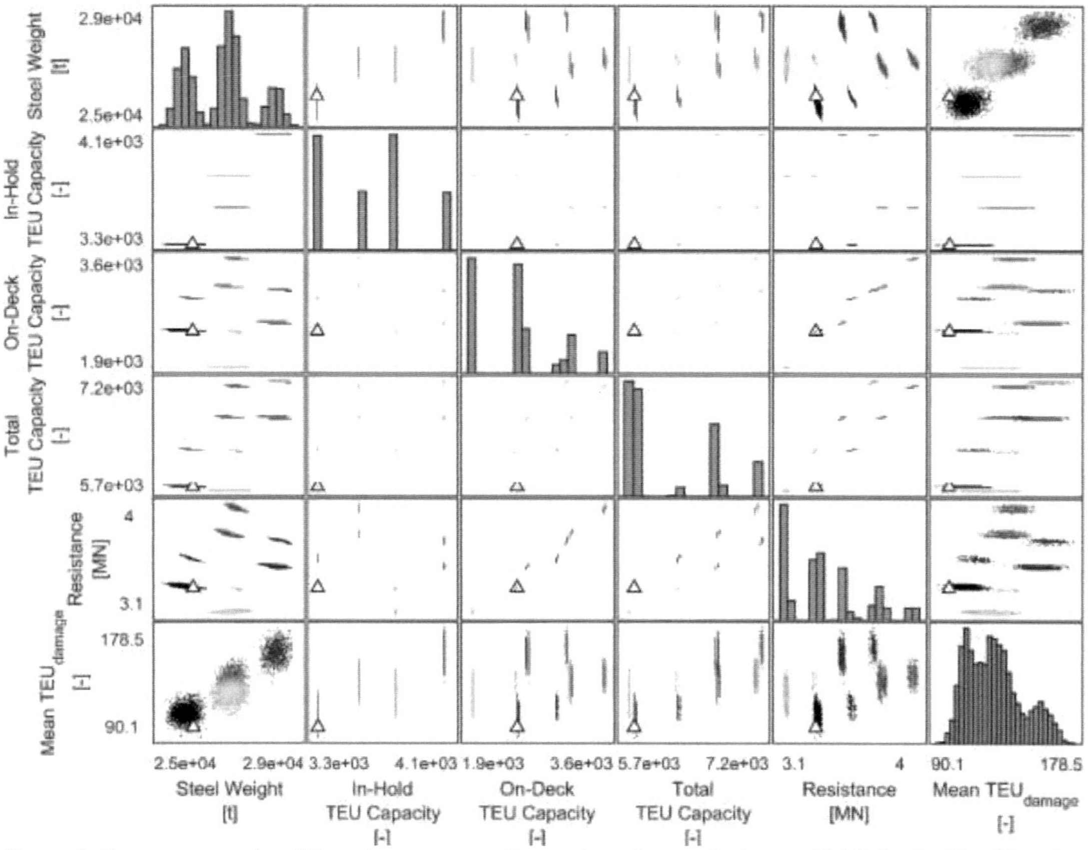

Figure 6: Output scatter plots (The points corresponding to the reference design are highlighted with white triangles; the upper and lower bounds of the data set are labeled along the axes. The four categories of ship size are distinguished by color coding: black for B = 15 rows and D = 9 tiers, red for B = 16 rows and D = 9 tiers, green for B = 15 rows and D = 15 rows and D = 10 tiers, blue for B = 16 rows and D = 10 tiers.)

VARIATION OF ASYMMETRIC SIDE HULL OF PENTAMARAN HULL RESISTANCE BY COMPUTATIONAL FLUID DYNAMICS (CFD)

Z Zikrina, Yanuar, and F A Nugroho, Universitas Indonesia, Indonesia

SUMMARY

Fast vessel technology is rapidly developing with the increasing of the demands in maritime field such as military, shipping and tourism. The substantial of designing fast vessel is to create a vessel with highest speed and lowest fuel consumption. Ship hull optimization is one of methods to reduce fuel consumption in designing the high-speed vessel. The advantages of multihull vessel are huge payload, wide deck area, good stability and seakeeping, also decrease the total resistance due to wave making resistance caused by wave interference of the hulls. Multi-hull vessel is divided by number of hulls which are catamaran, trimaran, quadramaran and pentamaran. In a research by Tuck (1998) shows that the wave resistance of quadramaran is relatively decreasing at the high speed. Thus, this matter tends to appear the curiosity regarding wave resistance of pentamaran at the high speed. Pentamaran is a kind of ship which has a main hull, two inner and outer side hulls. This research aims to show the configuration effect of pentamaran ship on resistance coefficient by conducting simulation at Froude Number 0.1-1. Computational Fluid Dynamics (CFD) method performed to predict the resistance of hull-form configuration. In this research, the calculation of pentamaran hull resistance is simulated by ANSYS – FLUENT (CFD Software Package). The multiphase region is separated by defining boundary air inlet as the flow of air and water inlet as the flow of water. The free surface model is performed in calm water condition by Volume of Fluid (VOF) model and k-epsilon is used to simulate the turbulent viscous model. The validation data obtained from the Computational Fluid Dynamics (CFD) method were found almost similar with latest pentamaran experimental data. There are two main components of resistance coefficient which dominate the value of total resistance coefficient on pentamaran ship, those are coefficient wave pressure resistance and coefficient viscous resistance. The result shows that waveform resistance coefficient has more contribution in reduce or raise the value of total coefficient resistance.

Keywords: fast vessel; multi-hull; pentamaran; ansys – fluent; volume of fluid

NOMENCLATURE

υ	Kinematic viscosity (N s m^{-2})
ρ	Density of water (kg m^{-3})
P	Pressure (N m^{-2})

1. INTRODUCTION

One of the essential thing in designing fast vessel is to create the hull model with the lowest resistance to reduce fuel consumption. Multihull vessels are categorized by the number of its hulls, those are catamaran, trimaran, quadramaran and pentamaran. Pentamaran or five hulls design is new innovation in multihull research to decrease fuel consumption. The pentamaran hull is able to produce very low resistance at high speed for certain hull configuration which is determined by configuration and form of each hull.

Pentamaran is going to be an interesting research topic to be discussed because it is possible to produce lower resistance than other forms of multi-hull at high speeds by changing the configuration of the side-hulls. The hull configuration of pentamaran varied based on two parameters, which are the ratio of the distance between the hull transversely and the length of the main-hull called as staggered (R/L), and comparison distance between the hull longitudinally to the length of the main-hull called as separation (S/L). Both of the parameters will be varied to be several configurations in order to generate the lowest value of resistance coefficient.

Investigation of pentamaran hull resistance has been performed in 2001 by Peng et al. Peng calculated the pentamaran ship resistances of two various configuration of staggered side hulls by numerical methods. The research has shown there are profits produced by interference of pentamaran at Froude Number 0.6 [1]. In 2013, Yanuar investigated the resistance of pentamaran ship by experimental methods. Those are nine pentamaran ship configuration varied by staggered and separation side hull arrangement. The result said that pentamaran ship with separation value 3/16 (S/L=3/16) has the lowest resistance [2]. Whereas, the staggered pentamaran ship configurations show uncertain value of coefficient resistance. In 1997, a research for asymmetric side hull on trimaran ship shown that it produces the lower resistance than the symmetric [3]. This paper will find out the effect of asymmetric side hull arrangement on pentamaran ship.

The Computational Fluid Dynamics (CFD) method has been succeeded in showing similar result with the experimental result. The simulation of water flow over the ship hull already implemented to determine ship resistance by using RANS Equation [4]. This paper presents variation of pentamaran hull configuration by varying the asymmetric hull arrangement and calculating the resistance by CFD software package, Ansys- Fluent.

2. THEORITICAL APPROACH

2.1 SHIP RESISTANCE

Ship resistance occurs caused by air and water interaction to the hull surface which produce losses wave energy. To analyse relation between ship resistance and speed, the coefficient resistance is compared to Froude Number in a Graph. The coefficient resistance is determine by Equation 1.

$$c = \frac{R}{\frac{1}{2}\rho V^2 S} \quad (1)$$

The Froude Number is determine by Equation 2.

$$Fn = \frac{V}{\sqrt{g.L}} \quad (2)$$

Viscous resistance and wave resistance are two main components of ship resistance as well as Molland's explanation (2008) as shown in Figure 1. The viscous resistance are supported by viscous form resistance and skin friction, whereas the wave resistance are effected by pressure resistance [5].

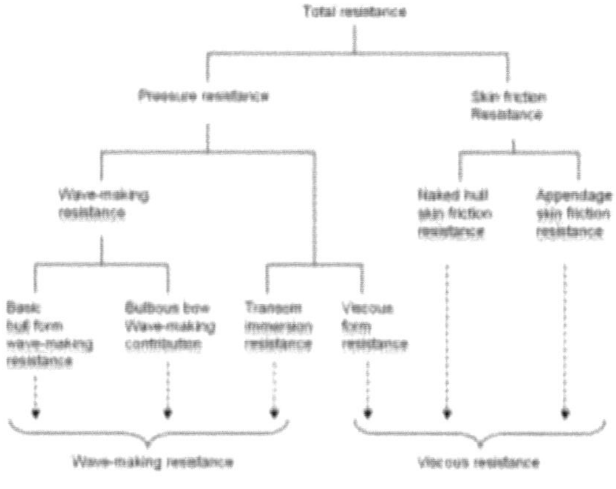

Figure 1. Ship Resistance Components [5]

The calculation results by Tuck (1998) shown the wave resistance has contribute to reduce the total resistance on multihull ship. Based on michell theory, the decresing of wave resistance caused by wave interaction produced by each hull [6]. In Figure 2 shows the characteristic of wave resistance to the Froude Number for catamaran, trimaran and quadramaran. Furthermore, a research regarding pentamaran hull resistance by Tarafder (2013) concludes that wave resistance coefficient relatively starts to decrease at Froude Number 0.8 [7].

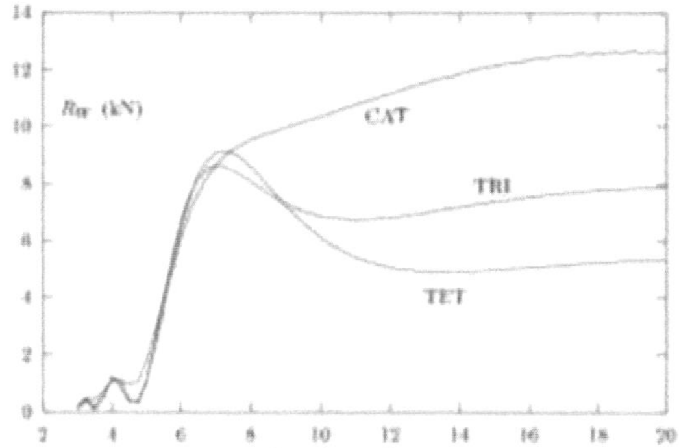

Figure 2. Wave Resistance Vs Speed [6]

2.2 ANSYS - FLUENT

Ansys – Fluent can be used to determine the compressible or incompressible, laminar or turbulent fluid flow and solve the problem based on momentum equation. In this case, the fluid flow is assumed as incompressible, viscous turbulent and multiphase flow. Hence, Reynold-Average Navier Stokes (RANS) is used to simulate the case. In a research by Sahoo [2015] about investigating trimaran resistance using CFD, Volume of Fluid (VOF) model is needed for modelling the free surface between air and water [8].

3. CFD SIMULATION

There are three main process of CFD Simulation, those are pre-processing, processing, and post processing.

- Pre-processing
  In this step, the geometry is prepared as the real condition and the boundary conditions are defined. Then, meshing process is built by define number and size of grid.
- Processing
  In this step, the computer is trying to solve the problem.
- Post-processing
  In this step, the result is prepared to be shown.

In this research, the CFD simulation started with design the model, defining meshing and configure the CFD setup.

3.1 MODELLING

Before starting the simulation, the essential part is preparing a model. Pentamaran has five hulls, which the main hull is located in the centre of body and others hull called side hull. Asymmetric hull forms are combined on the pentamaran ship as the side hulls. In this research, there are five pentamaran ship designs which varried by configure the asymmetric side hull arrangement as we can see at Figure 3.

- A : Asymmetric Outboard Outer side hull
- B: Asymmetric Inboard Outer side hull
- C: Asymmetric Outboard Inner side hull
- D: Asymmetric Inboard Inner side hull
- E: Symmetric

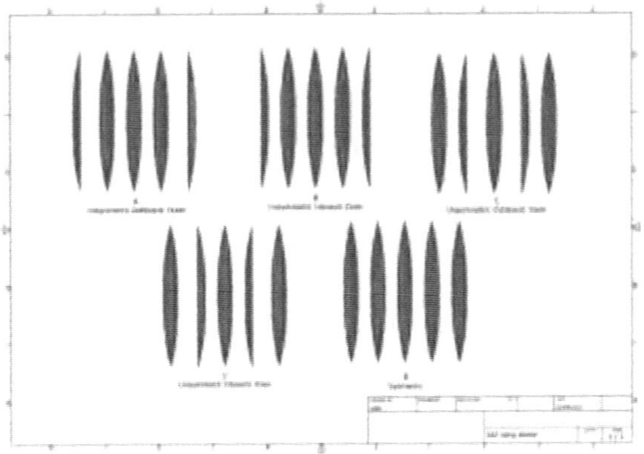

Figure 3. Variation of Pentamaran Ship

Each hull has the same hull form and dimension. The dimension of the hull shown at Table 1.

Tabel 1 Main Dimension

Main Dimension	
Lenght (m)	1.8
Beam (m)	0.18
Height (m)	0.17
Draft (m)	0.07

The wigley hull form has the lowest resistance of monohull. These pentamaran ship geometry are designed by lofted each section. The dimension of domain shown in Table 2.

Tabel 2 Domain

Domain				
Forward	0.5x	Loa	0.9	m
After	4x	Loa	7.2	m
Left Side	1x	Loa	1.8	m
Rigth Side	1x	Loa	1.8	m
Depth	2x	Loa	3.6	m
Upside	0.5x	Loa	0.9	m

The boundary condition is defined as air inlet & water inlet as velocity inlet, outlet as pressure outlet, walls and ship as walls, and top as symmetry are shown in Figure 4.

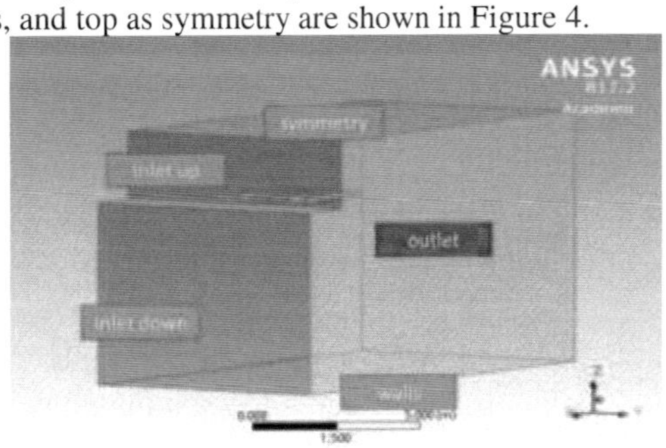

Figure 4 Boundary Condition

3.2 MESHING TECHNIQUE

Mesh quality and skewness are important parameters to create a good quality of meshing. If the value of mesh quality close to one, it describes a good meshing quality, and if the value is close to zero then the quality of meshing is poor. Contrary to the mesh quality, if the value of skewness close to one, it means a poor skewness of meshing and if the value close to zero means the skewness is fine.

In this section, meshing technique is needed to be implemented. There are many things affect mesh quality and skewness, such as size and type of elements. Therefore, the complexity of geometry is considered. Simple geometry is easy to create the structured grid meshing, then possible to decrees time setup. But, for the complex geometry, there will be special treatment to create a proper quality of meshing. There are five various meshing technique tried to create the best meshing grid. In this case, the geometry is quite complex, so that some special treatment are applied. The size of elements is 0.027, mesh quality is 0.99, and skewness is 0.0000041429.

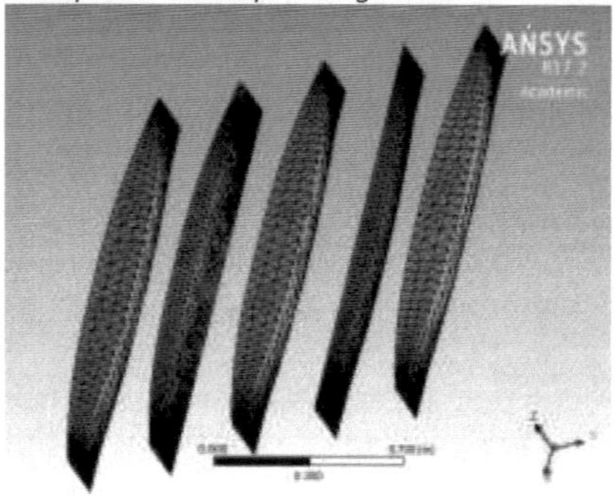

Figure 5 Meshing

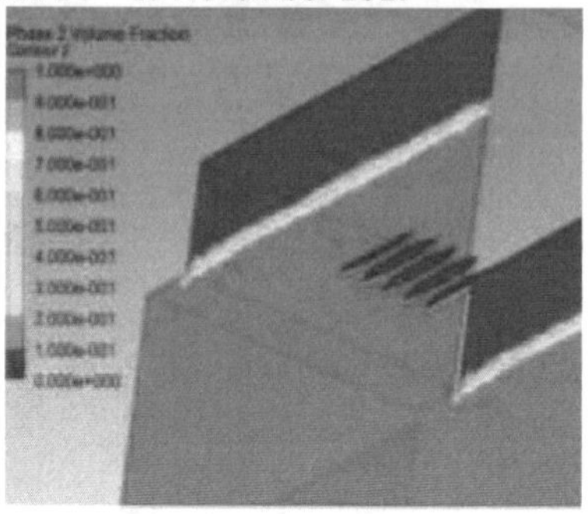

Figure 7. Volume Fraction

3.3 CFD SETUP

To determine the accuracy of CFD methods, validation step is needed by comparing between simulation and experimental result. The Figure 6 shows that the CFD methods performed a well-turned agreement for calculating ship resistance by Ansys- Fluent.

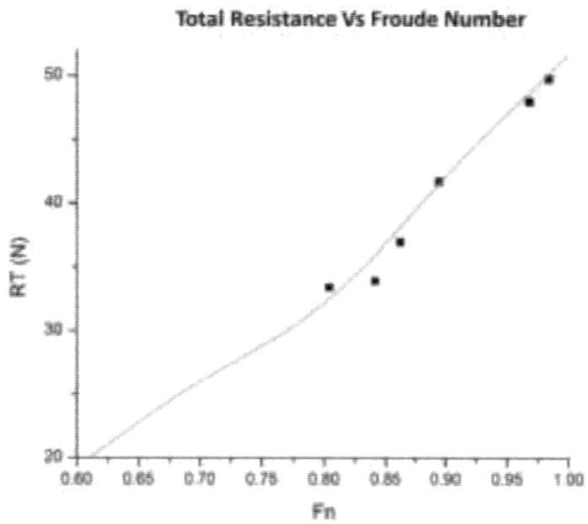

Figure 6 Validation

There are three types of multiphase model, those are Eularian, VOF (Volume of Fluid) and Mixture. In this case, VOF (Volume of Fluid) is used to modelling free surface by defining the value of surface tension between air and water. When petamaran ship is moving in a calm water, the fluid flows will become turbulence. Because the flow is assumed to be steady flow and constant velocity of ship, it is predicted that turbulent flow is not really complicated. Thereafter, k-epsilon turbulent model is applied to describe the turbulence phenomena in the simulation. Also, 2 % are given for the value of turbulent intensity and viscosity as. The volume fraction shown as Figure 7.

4. RESULT AND DISCUSSION

4.1 COMPONENTS OF PENTAMARAN SHIP RESISTANCE

4.1 (a) Asymmetric Outboard Outer side hull

The graph shows information regarding the characteristic components of resistance coefficient and the relation between its value and Froude Number on Pentamaran Configuration A. There are three kinds of curves shown, the black bold line illustrates the value of total resistance coefficient, the red strips illustrate the value of resistance coefficient of waveform and the blue dots represent the value of resistance coefficient of viscous.

It is clear that the total value of resistance coefficient is decreasing while the raising value of Froude Number. The resistance coefficient of waveform dominates the change of total value of resistance coefficient. Afterwards, the value of viscous resistance coefficient is relatively stable. It can be seen in Figure 8 that there is a slightly slump of the total value of resistance coefficient along Froude number 0.6 – 1. It is affected by the drop value of wave form resistance coefficient. It is suspected that this phenomenon occurs due to the interaction of waves between outboard asymmetric hulls which cause a decrease in the waveform resistance coefficient at high speed. In catamaran ship, the interaction between the asymmetric hulls is affected by the distance between each hulls and the rising of speed [9].

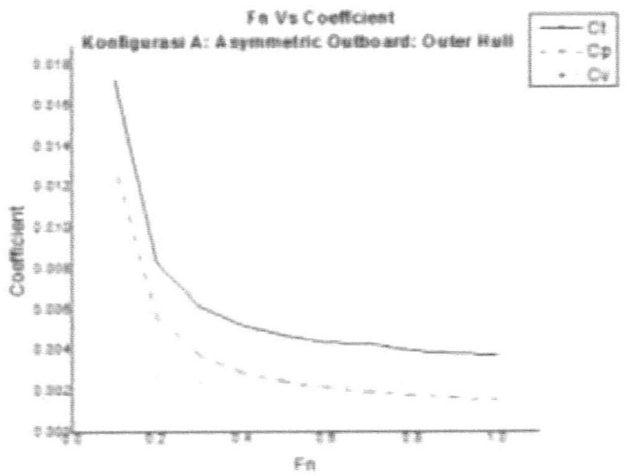

Figure 8 Asymmetric Outboard: Outer Hull

4.1 (b) Asymmetric Inboard Outer side hull

The graph of Pentamaran Configuration B illustrates the correlation between resistance coefficient and Froude Number. Three kind of curves represents total value of resistance coefficient (black line), value of waveform resistance coefficient (red strips), and value of viscous resistance coefficient (blue dots). Overall, the change of total resistance coefficient is similar with waveform resistance coefficient. This matter happened because the waveform resistance coefficient has a huge role in changing value of total resistance coefficient.

In the beginning, there is a significantly plung of total resistance coefficient along Froude Number 0.1-0.2. Whereas, the curve of waveform resistance coefficient is above the curve of viscous resistance coefficient. In range of Froude Number 0.2 – 0.7, there is a small decline of the total resistance coefficient since the decreasing of waveform resistance coefficient. The viscous value has no role to increase or decrease the total value of resistance coefficient. At Froude Number 0.8, there is a surge and it is predicted that the hump is affected by the dominance of wave resistance at high speeds [2]. This issue proves that actual wave resistance coefficients may contribute to raise or drop the total resistance coefficient [10]. In this case, the wave interaction generated by the hulls of the ship configuration which has two asymmetrical hulls on the inner side hull and outward, results in an adverse interaction as its contribution to the rise the waveform resistance coefficient value.

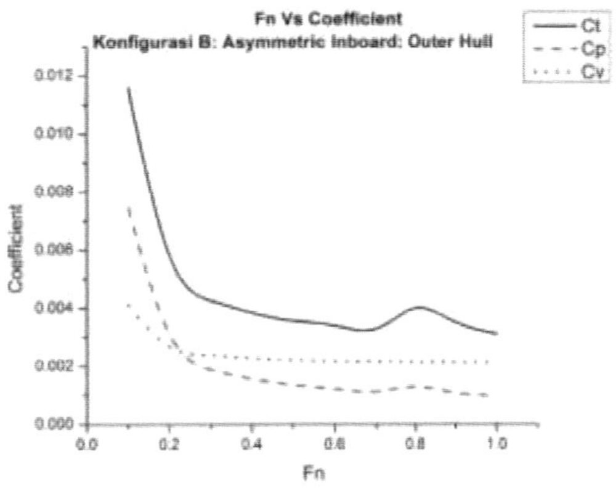

Figure 9 Asymmetric Inboard: Outer Hull

4.1 (c) Asymmetric Outboard Inner side hull

The chart compares the value of waveform resistance coefficient and viscous resistance coefficient to the change of Froude Number on Pentamaran Configuration C. The red strips show curve of waveform resistance coefficient, blue dots display curve of viscous resistance coefficient, also the black line expresses the total of those curves. As we can see that the value of waveform resistance coefficient is constantly above the curves of viscous resistance coefficient. As shown in Figure 10, the declining of total resistance coefficient is caused by the alteration of waveform resistance coefficient, while the curve of viscous resistance coefficient maintains the same level.

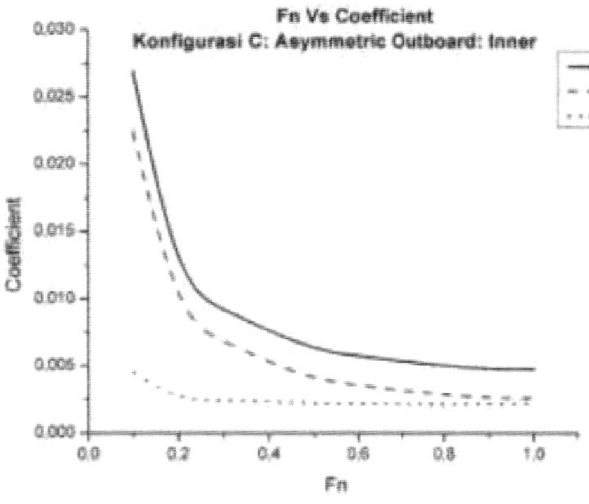

At first, the total resistance coefficient experiences a dramatically drop in the range of Froude Number 0.1-0.2. A moderate plummet of this also appears between Froude Number 0.2-0.4. Then, there is a slightly dive because of the change of waveform resistance coefficient. It might be expected that the outboard configuration of inner-hull asymmetric demonstrates the profit of wave interaction between each hull on pentamaran ship by reducing the value of waveform resistance coefficient at high speed.

4.2 (d) Asymmetric Inboard Inner side hull

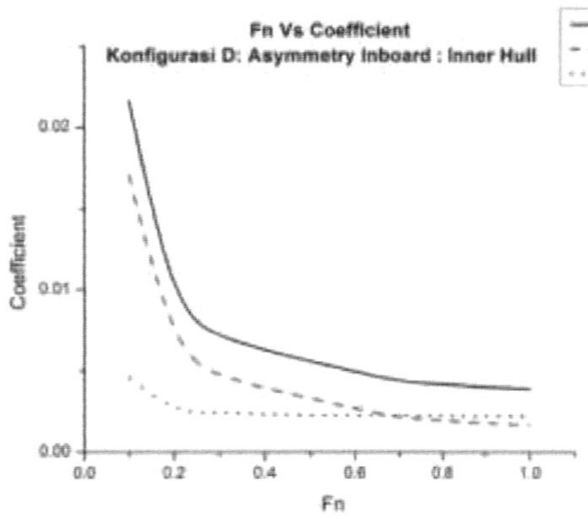

Figure 10. Asymmetric Inboard: Inner Hull

This graph concerns data regarding the different roles of waveform resistance coefficient and viscous resistance coefficient along Froude Number 0.1-1 on pentamaran configuration D. The black line shows the total of both kind of resistance coefficient, also the red strips illustrate the change of waveform resistance coefficient and the blue dots emphasizes the alteration of viscous resistance coefficient. Overall, the total resistance coefficient is relatively decreasing along the growing value of Froude Number.

An enormous decreasing value of total resistance coefficient lead off the change of it. While, the value of waveform resistance coefficient is regularly higher than value of viscous resistance coefficient along Froude Number 0.1-0.7. Following this, a slightly reduction occurs with the grow of Froude Number between 0.2-1. This matter exists due to the slum value of wave resistance coefficient. This evidence explains that the alteration of waveform resistance coefficient has a play role in changing value of total resistance coefficient of pentamaran at the high speed. This argument is supported by stable value of viscous resistance coefficient between Froude Number 0.2-1.

4.3 (e) Symmetric

The chart presents data concerning the contribution between waveform resistance coefficient and viscous resistance coefficient on symmetric pentamaran. The black line is displayed as total resistance coefficient, red strips demonstrate waveform resistance coefficient and the blue dots is represented as viscous resistance coefficient. The viscous resistance coefficient has small role in affecting the value of total resistance coefficient.

In the beginning, total resistance coefficient experienced a tremendously fall in the range of Froude Number 0.1-0.2. Following this, the value of waveform resistance coefficient is eternally higher than value of viscous resistance coefficient for Froude number 0.1-0.6. At Froude Number 0.2-1, there is a fluctuation of total resistance coefficient in accordance with the change of wave resistance coefficient.

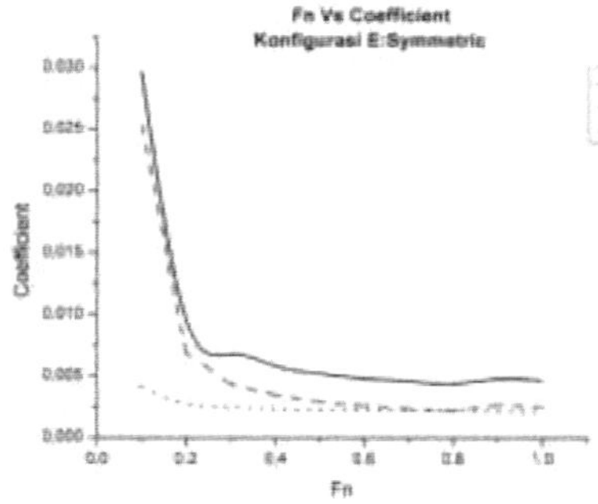

Figure 11. Symmetric

4.3 Total Resistance Coefficient

The figure 12 explain comparison data regarding total resistance coefficient of Pentamaran. There are five configuration of pentamaran hull which is distinguished the arrangement and form of asymmetric side-hull. These curve has similar likelyhood trendline which has a dramatically decrease at early Froude Number and relatively small decrease at high speed. For exceptional, the total resistance coefficient of symmetric pentamran has small fluctuation value at the high speed.

In the beginning, Pentamaran E or symmertic pentamaran leads the value of total resistance coefficient at Froude Number 0.1. It can be seen that Pentamaran B has the lowest value of total resistance coefficient along Froude Number 0.1-1. While, the highest value of total resistance coefficient shown by Pentamaran C for Froude Number 0.2-1.

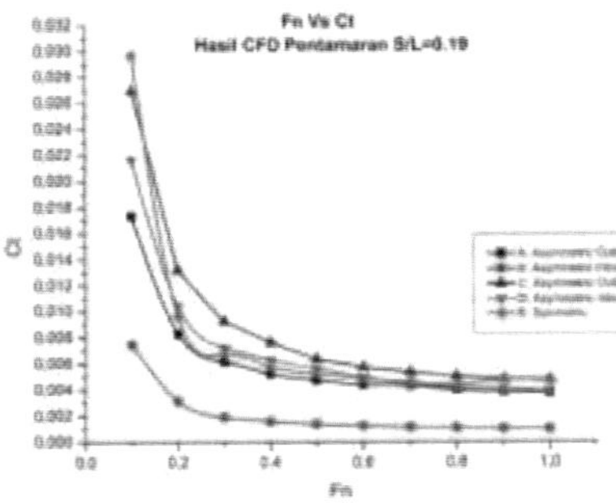

Figure 12. Interference Factor

7. CONCLUSIONS

After investigation, there are two major components affected the value of pentamaran resistance, which are coefficient of waveform and viscous. It is clear that the waveform coefficient has a great contribution to raise or decline the value of total resistance coefficient. In this research, Pentamaran B which has Inboard Outer asymmetric hull as its side hull is recommended to be learnt futher in order to its implementation as pentamaran fast vessel. Since, Pentamaran B has the lowest total resistance coefficient caused by a profitable wave interation.

8. ACKNOWLEDGEMENTS

The authors would like to express sincere gratitude to Univeristas Indonesia to support our research.

9. REFERENCES

1. Peng, Hongxuan, 2001. Numerical computation of multi-hull ship resistance and motion, DALHOUSIE UNIVERSITY Halifax.

2. Yanuar, Gunawan, Talahatu, Indrawati and Jamaluddin. Resistance analysis of unsymmetrical trimaran model with outboard sidehulls configuration. Scopus. Volume 12, Issue 3, September 2013, Pages 293-297

3. Ackers BB, Michael TJ, Tredennick OW, Landen HC, Miller III ER., Sodowsky JP, Hadler JB (1997). An Investigations of the Resistance Characteristics of Powered Trimaran Outriggers Configurations. SNAME Trans. 105, 349-373.

4. Chang Hwan Son. CFD Investigation of Resistance of High-Speed Trimaran Hull Forms. The College of Engineering at Florida Institute of Technology. 2001.

5. Molland, A.F., Turnock, S.R., dan Hudson, D.A. (2011), Ship Resistance and Propulsion: Practical Estimation of Ship Propulsive Power, Cambridge University Press, New York, USA.

6. Tuck EO, Lazauskas L, 1998. Optimum spacing of a family of multi-hull. Ship Technology Research, 45, 180-195.

7. Md. Shahjada Tarafdera, Mir Tareque Alia, Md. Shahriar Nizamb. Numerical prediction of wave-making resistance of pentamaran in unbounded water using a surface panel method. 2013. Bangladesh.

8. ANSYS FLUENT. (2013), "ANSYS FLUENT Theory Guide" ,ANSYS Inc, U.S.A

9. R. Broglia, B. Jacob, S. Zaghi, F. Stern, and A. Olivieri, "Experimental investigation of interference effects for high-speed catamarans," Ocean Engineering, vol. 76, pp. 75–85, 1/15/2014.

10. Chang Hwan Son. CFD Investigation of Resistance of High-Speed Trimaran Hull Forms. The College of Engineering at Florida Institute of Technology. 2001.

9. AUTHORS BIOGRAPHY

[Zulfah Zikrina] holds bachelor degree of Naval Engineering at Universitas Indonesia. She is responsible for CFD (Computational Fluid Dynamics) methods.

[Firman Ady Nugroho] holds lecturer as current position at Universitas Indonesia. He is responsible for numerical methods. His previous experience includes ship structure and welding.

[Yanuar] holds the current position of professor at Universitas Indonesia. He is responsible for experimental data. His previous experience includes multihull ship resistance by experimental methods.

IN-SITU TURBULENT BOUNDARY LAYER MEASUREMENTS OVER FRESHLY CLEANED SHIP-HULL UNDER STEADY CRUISING

B Nugroho, University of Southampton, Southampton, UK
R Baidya, University of Melbourne, Melbourne, Australia
M N Nurrohman, Institut Teknologi Sepuluh Nopember, Surabaya, Indonesia
A K Yusim, Universitas Jember, Jember, Indonesia
F A Prasetyo, Biro Klasifikasi Indonesia, Jakarta, Indonesia
M Yusuf, PT Dharma Lautan Utama, Surabaya, Indonesia
I K Suastika, Institut Teknologi Sepuluh Nopember, Surabaya, Indonesia
I K A P Utama, Institut Teknologi Sepuluh Nopember, Surabaya, Indonesia
J P Monty, University of Melbourne, Melbourne, Australia
N Hutchins, University of Melbourne, Melbourne, Australia
B Ganapathisubramani, University of Southampton, Southampton, UK

SUMMARY

A novel study on assessing the drag penalty due to hull roughness from a recently cleaned and painted ship hull is reported. Here we estimate the rough surface drag penalty by measuring the velocity profile directly over the hull of an operating ship under steady cruising using a Laser Doppler Anemometer (LDA). The novel technique is non-intrusive and requires only optical access and does not require a full hull-penetration. For this experiment a small glass window is placed on the double-bottom hull of an operating ship, allowing the LDA to measure the velocity gradient in the turbulent boundary layer formed over the hull (during steady sailing) across some traversable distance from close to the hull surface, to at least the end of the logarithmic region. Preliminary initial results show that there is an approximate 37% increase in skin-friction drag for a recently cleaned ship-hull compared to the hydrodynamically smooth surface.

This document is formatted in the convention required for all conference papers

NOMENCLATURE

[Symbol]	[Definition] [(Unit)]
ν	Kinematic viscosity (N s m^{-2})
U	Mean velocity (m s^{-1})
U_τ	Skin friction velocity (m s^{-1})
U_∞	Free stream velocity (m s^{-1})
L	Characteristic length scale (m)
δ	Boundary layer thickness (m)
z	Wall normal distance (m)
κ	Karman constant
A	wall intercept
Re_x	Reynolds number
C_f	Coefficient of friction
δ	Boundary layer thickness (m)

1. INTRODUCTION

Skin-friction drag that arises from turbulent boundary layers formed over the ship hull is one of the primary sources of energy consumption in the maritime sector. Lackenby [1] and Kodama [2] estimate that around 80%- 90% of the total drag experienced by a large bulk carrier is due to skin-friction drag. If we combine it with the fact that the shipping industry is one of the major energy users in the world, with estimated 200–300 million metric tonnes of fuel consumed by more than 100,000 ships annually [3-5], it is easy to see that the a large proportion of the industry's fuel usage is to overcome the skin-friction drag. Furthermore, the oil that these ships burn is mostly of a low grade, with sulphur content that can be many times more than is permitted in diesel fuel. Hence, beyond the large energy and economic footprint, skin-friction drag also directly contributes to air pollution and health problem [3-6].

The negative energy, health, environmental, and economic impacts of skin-friction drag is further exacerbated by the presence of surface roughness [2, 7- 18]. Hull surface roughness is an important, but largely unquantifiable contributor to the overall energy expenditure, emissions, and pollution from the shipping industry. These hull roughnesses are generally in the form of mechanical defects or biofoulings (the settlement of marine organisms on a surface in an aquatic environment). From a non-hydrodynamic perspective, a ship hull can often seem relatively smooth, particularly when it has just recently experienced dry-dock, where the hull is cleaned, sand-blasted, water-blasted, and repainted with anti-corrosion and anti-fouling paints. However, close inspections reveal that the hull can exhibit an "orange peel" roughness pattern that is above the ideal smooth state. Reports by Schultz [13,14] show that a plate in towing tank which is covered by biofouling and later cleaned, exhibit a higher skin-friction drag than a fresh new plate. The higher drag may arise from defects that come from the cleaning process. For a real case scenario, such as operating ship, the hull imperfection from repeated cleaning and painting process may be even more severe.

Considering the issues caused by ship hull roughness, it is imperative for naval architects and fluid dynamicists to characterise it and estimate the full-scale ship drag penalty. This would allow ship operators to estimate fuel usage and schedule the appropriate time for cleaning. Currently, the most accurate methods in estimating the drag penalty from ship hull roughness are via laboratory replication and experiment [8-18]. This method generally involves imprint and scanning of the hull roughness. The digital data is then scaled (to match the laboratory's Reynolds number) and replicated via CNC, 3D-Printing, or casting. The replicated roughness is then laid inside a wind tunnel, water tunnel, or towing tank and the flow over the roughness is measured via hot-wire anemometer, LDA, force balance, or Particle Image Velocimetry (PIV). Some studies use rotating disk or rotating cylinder apparatus to characterise the roughness (see Lindholdt et al [17] for further review). The main issue with these techniques is the cost, both in time and laboratory facility [17-19]. Furthermore, the scanning from the actual hull-roughness is generally performed over a small area, and we assume that it represents the entire ship hull (i.e. the entire hull surface roughness is homogeneous). Hence the laboratory measurement may not be able to capture the actual roughness surface flow over the hull. Another major challenge in performing laboratory experiment is the difficulty to match a real-world Reynolds number (i.e. flow over ship hull).

One method to overcome the laboratory experiment issue is to perform the measurement in-situ over a ship hull. This type of measurement technique generally uses Pitot tube that is attached under the hull of operating ship and [20-22]. The Pitot tube can be in a form of a Pitot tube rake [21] or a single Pitot tube attached to a traverse system that can move vertically in the hull-normal direction z [20,22]. These techniques allow one to measure the velocity profile over ship hull and estimate the skin friction velocity U_τ via the logarithmic -law of the wall. Lewthwaite et al [22] measured the mean velocity of a 23 m long ship hull over a two year period to monitor the effect of biofouling on the skin friction drag. Their result reveals an approximately 80% increase in skin friction which corresponds to a 15% loss in ship velocity at a matched fuel usage. Although this technique allows Lewthwaite et al [22] to measure the mean velocity U, the use of a manometer prevented them from obtaining the turbulence intensity or other higher order statistics. Furthermore, the use of a large Pitot tube may disturb the flow due to its intrusive nature and is also prone to blockage by fouling.

In order to obtain higher statistical reading without perturbing the flow over the ship hull, it is necessary to use a high-speed non-intrusive measurement technique. One such technique is via Laser Doppler Anemometer (LDA). In this study we report preliminary measurements with an LDA where we measure the mean velocity profile over a recently cleaned and painted ship hull under sailing conditions.

2. EXPERIMENT SET-UP

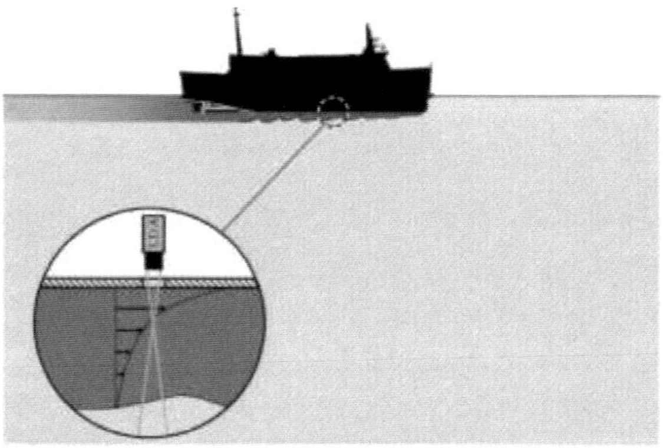

Figure 1: Illustration of the LDA set-up that measure the mean velocity profile inside the turbulent boundary layer that develop over ship's hull.

The experiment here is to use an LDA that is looking outwards via a small window/optical access installed in the bottom of the hull of a ship. The LDA is attached to a traverse system that measures the mean velocity profile inside the turbulent boundary layer (figure 1).

2.1 TEST BED SHIP DETAILS

Figure 2 : Dharma Kencana IX ferry from PT Dharma Lautan Utama (PT DLU)

The ship that is used for the experiment trial is a 70 m Roll-on/roll-off ferry Dharma Kencana IX that belong to PT Dharma Lautan Utama (figure 2). The ship operates daily serving Merak-Bakauheni line in Sunda Strait, Indonesia. The route is the busiest in Indonesia connecting the island of Java and Sumatera. The ship cruising velocity is around 9-10 knots (depending on weather), which translates to approximately 4.5 – 5 m/s. The constant velocity cruise time between the two islands is approximately one hour and fifteen minutes. The regular route allows us to perform repeatable measurement over a relatively consistent environment.

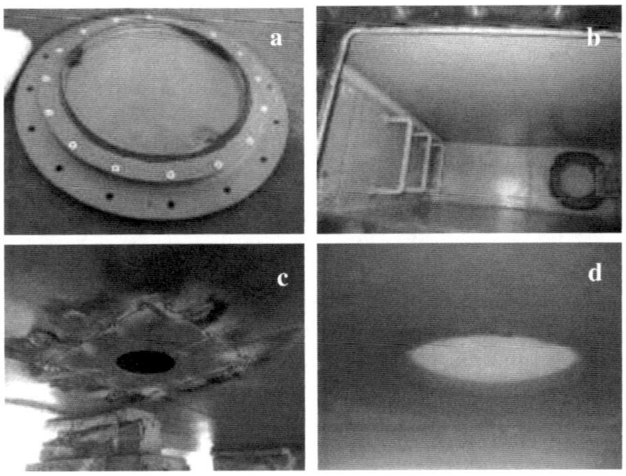

Figure 3: The water tight enclosure and window access at the double bottom hulls. (a) The optical access frame and glass, (b) Water tight enclosure inside double bottom hull with the optical access attached, (c) The circular cut for the optical access viewed from outside during dry docking, (d) The optical access viewed from underwater outside the ship's hull

A window was installed on the underside of the hull, located approximately 25.5 m downstream of the bow of the ship during its annual dry-docking and hull cleaning. The window is enclosed within a water-tight enclosure that is constructed between the double bottom hulls which also houses the LDA and computer controlled traversing rail. This enclosure ensures the safety of the ship in the unlikely event of a window failure. Figure 3 shows the enclosure and window access at the double bottom hulls. The window has a diameter of 300 mm and it is made of two tempered glass discs (with thickness of 10 mm and 12 mm) laminated with Polyvinyl butyral/PVB (with thickness of 1.52 mm) to ensure integrity.

2.2 HULL CONDITION

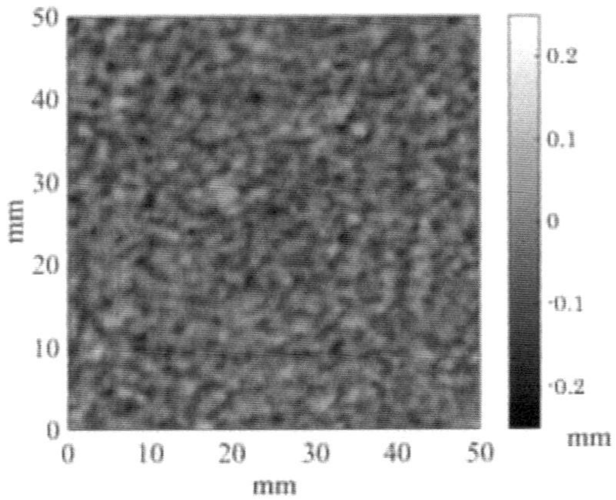

Figure 4: Lase scanner scan from a recently cleaned and painted ship hull.

During the annual dry docking, the Dharma Kencana IX ship hull is cleaned via sand and water blasting and then protected with anti-fouling paint under the supervision of anti-fouling paint producer. Figure 4 shows the laser scanning result from the "clean-hull". It has an "orange peel" pattern with roughness of ranging from 0.1-0.5 mm. Based on our personal communication with anti-fouling producer representatives, this type of hull-roughness can be found in many recently dry-docked ships, and it is relatively common. A much more severe "orange peel" pattern is even regularly found on many recently dry –docked ships.

2.3 LDA SYSTEM

The LDA system used is a Dantec two-component FlowExplorer Laser Doppler Anemometer that is attached to a Velmex computer controlled traverse system (see figure 5). The LDA consists of two cross beams at the wavelength range of 650 – 670 nm (red) and 770 – 810 nm (near infra-red). The LDA is only used to acquire the instantaneous velocity in the boundary layer when the ship is operating at cruise speed and hence maintaining a constant free stream velocity.

Figure 5: LDA system set up inside the water tight enclosure.

.

In this experiment no artificial seeding in the flow is introduced, instead, the LDA relies on the natural small particles in the sea-water. The use of artificial seeding, such as glass particles is a safety hazard and can contaminate the port and sea, endangering the marine life and human alike, particularly the traditional fisherman. In this LDA experiment we use a lens with a standoff range of 500 mm.

2.4 FREE STREAM VELOCITY

Figure 6 shows the free stream velocity/cruise velocity of the ship during the LDA data acquisition. The velocities are obtained through the on board GPS system. The data shows that the ship's velocity is approximately 4.5 m/s. Note that there are slight velocity variations during data acquisition. This is due to external effects such as wind, current, and wave. The free stream velocity variation will affect the hull's mean velocity profile measured by the LDA. However, for this type of field study, this variation is considered acceptable.

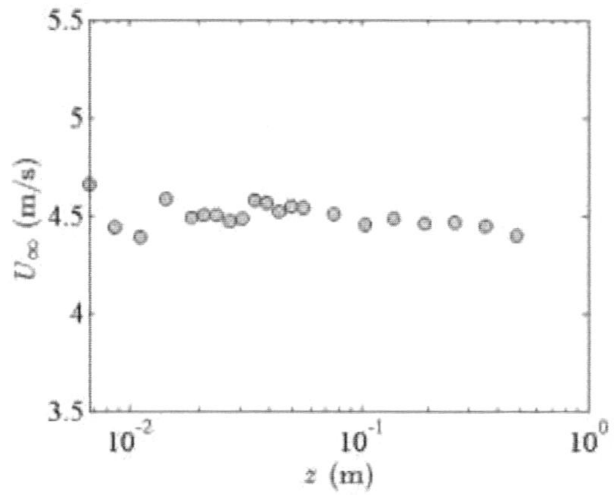

Figure 6: Free stream velocity/cruise velocity measured for each wall-normal position

3. RESULTS AND DISCUSSIONS

3.1 MEAN VELOCITY PROFILE

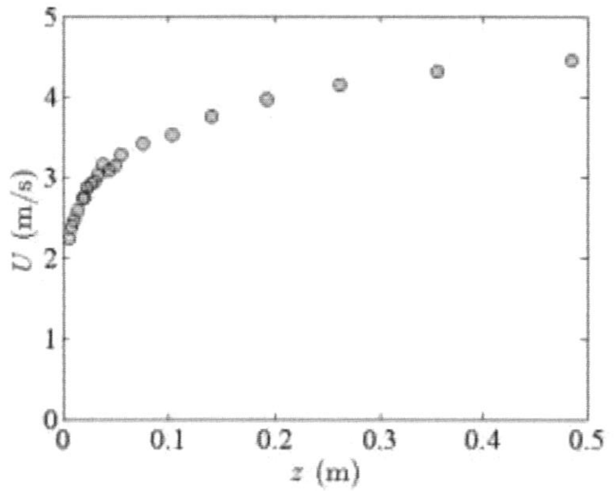

Figure 7: Mean velocity profile of the LDA reading, where U is mean velocity profile and z is wall normal distance.

Figure 7 shows the mean velocity profile of the LDA measurement over the ship hull. Here the LDA measures from a hull normal position of 6.7 mm to 485 mm with 21 logarithmically spaced positions. In this experiment, the initial measurement is located relatively far from the wall due to the difficulty in measuring the near wall flow. This is mostly due to the wall presence that changes the laser optical path and causes a shift in the measurement volume [23]. This issue is exacerbated by inconsistency in the size of natural seeding. LDA works by detecting Doppler shift from the laser that is scattered by particles (seeding) that move with the flow. The seeding particles are generally in the order of microns [24]. However, due to the lack of artificial seeding with constant size this may lead to difficulty in detecting instantaneous velocity.

The LDA starts to lose its data rate reading as we traverse further from the surface and approach the edge of the boundary layer. This may due to the laser wavelengths (red and near infra-red) that are severely attenuated in water. Hence, this has prevented us from measuring the flow velocity further from the wall. Furthermore, here we are also limited to use only 500 mm lens, as other longer range lenses cause further reductions in the obtainable data rate. Despite these issues, we have managed to acquire enough data samples to obtain the mean velocity profile.

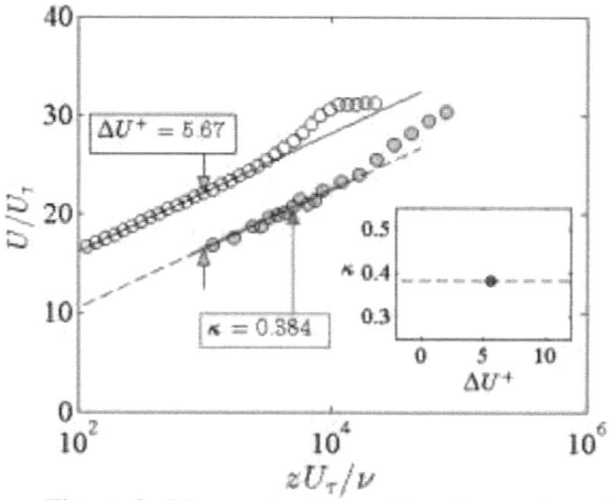

Figure 8: Mean velocity profile of smooth wall (from Marusic et al [27]), marked with open circles; and the rough ship hull, marked with grey circles. The inset figure shows the value of Hama roughness function against κ.

From the mean velocity profile, we are able to determine the skin friction velocity U_τ via a modified Clauser method [25, 26], by fitting the mean velocity profile to the logarithmic law:

$$\frac{U}{U_\tau} = \frac{1}{\kappa} \log\left(\frac{z U_\tau}{\nu}\right) + A - \Delta\left(\frac{U}{U_\tau}\right) \quad (1)$$

where κ = 0.384 is the Karman constant, A = 4.2 is the wall intercept, and ν is kinematic viscosity of water. Due to the non-smooth nature of the hull surface, the entire profile will be shifted downwards by the Hama roughness function $\Delta(U/U_\tau)$. In the smooth wall case, the Hama roughness function is zero, i.e. no vertical shift in the mean velocity profile.

Figure 8 shows the mean velocity profile of a smooth wall reference (open circles), taken from Marusic et al [27]; and the rough ship hull measurement from the LDA reading (grey circles). The straight dark line is the log law over the smooth wall with Hama roughness equal to zero, and the dashed black line is the logarithmic law over the rough surface (equation 1). The figure clearly shows the a downward shift of the LDA measurement when compared to the smooth wall, with Hama roughness function $\Delta(U/U_\tau)$ around 5.67, which indicates an increase in drag penalty.

3.2 INCREASE OF SKIN FRICTION COEFFICIENT C_f

Using the mean velocity profile of turbulent boundary layer over the smooth and rough wall (see figure 8), and the mean momentum integral equation, the local skin friction coefficient C_f as a function of distance over a flat plate can be obtained. Here we follow the methods of Monty et al [18] m to estimate it numerically. Figure 9 shows the predicted average skin friction coefficient for the smooth surface (open circles) and the rough hull (grey circles) for ranges of Reynolds number,

$$Re_x = U_\infty L/\nu \quad . \qquad (2)$$

Here U_∞ is the ship free stream velocity/cruising velocity, L is characteristic length scale (or in here represents the distance from the front of the ship to the optical access). The vertical dashed black line represents the ship hull at a downstream location of 25.5 m, where the optical access located, and a cruise speed of 5 m/s.

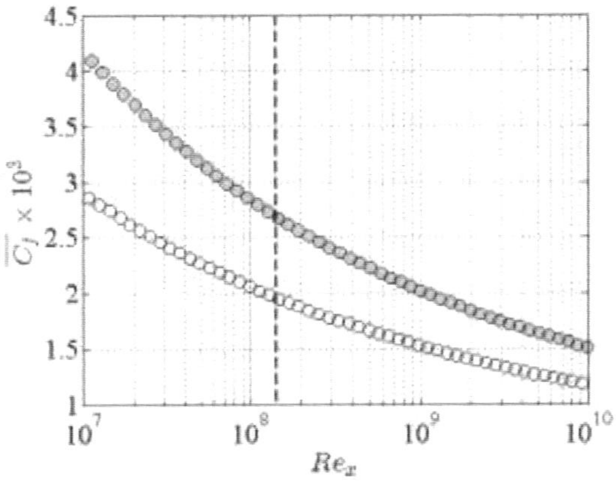

Figure 9: Average skin friction coefficient against Reynolds number for the (black circles) smooth surface and (grey circles) for the rough surface. The vertical dashed black line is the Reynolds number of the ship at downstream location of 25.5 m (optical access location) and cruise speed of 5 m/s.

From figure 9, the coefficient of friction for the hydrodynamically smooth wall at $Re_x = 1.4 \times 10^8$ is 1.96 x 10^{-3} (at the cross of vertical dashed line and open circle), while for the rough surface the coefficient of friction is 2.69 x 10^{-3} (at the cross of vertical dashed line and grey circle). From this calculation, we found that there is an increase of drag by almost 37% when it is compared to a hydrodynamically smooth wall. Such an increase is relatively significant. This result shows that even a freshly cleaned and painted ship hull, may already suffer from 37% increase of drag penalty.

Note however, that the result still needs further analysis and validation using other LDA system. Our LDA system uses red and infra-red lasers that suffer from attenuation when it measures velocity of liquid (sea water). Hence there is a possibility that we did not capture the entire flow dynamics. Further studies using an LDA that uses a blue and green laser, which will suffer less from attenuation would be desirable and would allow us to obtain a higher data rate, and hence obtain additional turbulence statistics.

3.3 POSSIBLE EFFECT FROM ROUGHNESS HETEROGENEITY AND WELDING SEAM

Beyond the "orange peel" pattern such as in figure 4, there are other possible causes of the drag penalty. Two of which that we believe have the highest impacts are biofouling heterogeneity and welding seam.

Welding seam is a common feature in ship hull that connects panels that form the hull profile. Many of these seams have significant thickness, around 2 - 10 mm. These features can act as additional trips that cause internal boundary layers that affect our reading. Moreover, it may also act as a blockage in the flow. Figure 10 shows one such seam that is located a few meters upstream of our optical access. Removing these seams would result in additional cost for the ship manufacturer and operator, hence they are rarely removed, but will reduce drag which may eventually yield net savings over the operating life of the ship.

Figure 10: Biofoulings from previous dry-docking cycle that were not cleaned due to inaccessible caused by the docking blocks and welding seam.

The drag penalty experienced by mechanical defects on ship hull is further exacerbated by the roughness heterogeneity. This heterogeneity is mostly caused by the patches of biofouling that remain from the previous cycle of dry-docking. These inaccessible patches are mostly caused by the docking blocks (see figure 10). From rivate communication with ship operator representatives and anti-fouling producers, it is very common to have such heterogeneity. This is due to the high cost of the dry-docking process. Removing the biofouling that is covered by the docking blocks will require an extra cycle of an expensive dry docking process (requiring a refloat).

Such heterogeneity and additional `trips' due to weld seams are believed to affect the ship's skin friction drag. Our LDA measurement may also pick up the secondary flow and internal boundary layer that is generated by the biofouling heterogeneity and welding seam located upstream of the optical access. This could also shift the Hama roughness function lower in the mean velocity profile, resulting in higher drag penalty. In the future the upstream fetch between the optical access and the front of the ship will be more thoroughly surveyed, and monitored during the experiment.

Persuading ship operators to remove such welding seam and biofouling heterogeneity would require a good economic reason. In order to provide a better argument, it is desirable to perform another measurement on the same ship where its welding seam and biofouling heterogeneity are removed. This would reduce the additional drag penalty and improve ship's efficiency. From the turbulent flow study perspective, this will also provide us with a more controlled experiment.

4. CONCLUSIONS

A study of in-situ turbulent boundary layer measurement from a recently cleaned and painted ship hull via LDA is reported. The result from this investigation shows that even a "clean" hull may already suffer from a substantial drag penalty increase compared to the theoretically ideal hydrodynamically smooth surface. Our data shows that the increase may up to 37%. The study shows the sobering challenges facing the shipping industry in dealing with surface roughness.

The results presented here reflect a preliminary investigation. Further studies using a more powerful LDA (with blue and green laser) are needed to provide better turbulence statistics. Furthermore, this also would allow us to measure further from the wall, up to boundary layer thickness δ, increasing the accuracy with which local C_f can be determined

The hull roughness drag penalty is exacerbated by the welding seam and heterogeneity that arises from biofouling that was not cleaned in the previous dry- docking cycle (due to coverage by the dock blocks). Hence there is a need to investigate this type of roughness in more detail. This includes a direct drag penalty comparison between a ship that has welding seam and biofouling heterogeneity, and the same ship that has those roughness removed.

5. ACKNOWLEDGEMENTS

The authors would like to thank the Australian Research Council (ARC) and the Newton Fund from the British Council for financial support.

6. REFERENCES

1. LACKENBY, H., 'Resistance of Ships, With Special Reference to Skin Friction and Hull Surface Condition. Proceedings of the Institution of Mechanical Engineers, 176 :981– 1014, 1962.

2. KODAMA, Y., KAKUGAWA, A., TAKAHASHI, T., and KAWASHIMA, H,. "Experimental study on microbubbles and their applicability to ships for skin friction reduction", Int. J. Heat and Fluid Flow, Vol.21, pp.582–588, 2000.

3. CORBETT, J. J., WINEBRAKE, J. J., GREEN, E. H., KASIBHATLA, P., ERYING, V., LAUER, A., Mortality from ship emissions: A global assessment. *Environ. Sci. Technol.,* 41(24) : 8512 – 8518, 2007.

4. WINEBRAKE, J. J., CORBETT, J. J., GREEN, E. H., LAUER, A., ERYING, V., Mitigating the health impacts of pollution from oceangoing shipping : An assessment of low-sulfur fuel mandates. *Environ. Sci. Technol.* 43 (13) 4776 – 4782, 2009.

5. ERYING, V., ISAKSEN, I., BEMSTEN, T., COLLINS, W. J., CORBETT, J. J., ENDERSEN, O., GRAINGER, R. G., MOLDAVONA, J., SCHLAGER, H., STEVENSON, D. S., Transport impacts on atmosphere and climate:

Shipping. *Atmos. Environ.*, 44(37):4735–4771, 2010.

6. MOLLAND, A. F., TURNOCK, S. R., HUDSON, D. A. and UTAMA, I. K. A. P. (2014) Reducing ship emissions: a review of potential practical improvements in the propulsive efficiency of future ships. Transactions of Royal Institution of Naval Architects Part A. 156, 175-188

7. NIKURADSE, J. Gesetzmassigkeiten der turbulenten stromung in glatten rohren. *Forsch. Auf Dem Gebiet des Ingen.*,3:1–36, 1932

8. GRANVILLE, P.S. The frictional resistance and turbulent boundary layer of rough surfaces. J. Ship. Res 2: 52–74, 1958.

9. GRANVILLE, P.S. Three indirect methods for the drag characterisation of arbitrarily rough

10. surfaces on flat plates. J. Ship. Res 31(1): 70– 77, 1987.

11. GRIGSON, C. The full scale viscous drag of actual ship surfaces and the effect of quality roughness on predicted power. J. Ship. Res 31(3):189–206, 1987.

12. GRIGSON, C. Drag losses of new ships caused by hull finish. J. Ship. Res 36(2):182–196, 1992

13. JIMENEZ, J.,Turbulent flows over rough walls. *Annu. Rev.Fluid Mech.*, 36:173–196, 2004.

14. SCHULTZ, M. P., (2004) Frictional resistance of antifouling coating systems. *J. Fluids Eng.* 126(1039), 2004.

15. SCHULTZ, M. P., Effects of coating roughness and biofouling on ship resistance and powering. *Biofouling* 23(5-6):331–341, 2007

16. SCHULTZ, M. P., BENDICK, J. A., HOLM, E.R. & HERTEL, W. M., Economic impact ofbiofouling on a naval surface ship. *Biofouling* 27: 87–98, 2011.

17. FLACK, K. A., SCHULTZ, M.P., 2010. Review of hydraulic roughness scales in the fully rough regime. J. Fluids Eng. 132, 041203.

18. LINDHODLT, A., DAM-JOHANSEN, K., OLSEN, S. M., YEBRA, D. M., KIIL, S., 2015. Effects of biofouling development on drag forces of hull coatings for ocean-going ships: a review. *J. Coat. Technol. Res.* 12(3) 415-444

19. MONTY, J. P., DOGAN, E., HANSON, R., SCARDINO, A. J., GANAPATHISUBRAMANI, B., HUTCHINS, N., An assessment of the ship drag penalty arising from light calcareous tubeworm fouling. *Biofouling*, 32(4): 451-64, 2016.

20. NUGROHO, B, UTAMA, I. K. A. P., SUASTIKA, I.K., PRASETYO, F. A., YUSUF, M., TULLBERG, M., MONTY, J. P., HUTCHINS, N., GANAPATHISUBRAMANI, B., Managing international collaborative research between academics, industries, and policy makers in understanding the effects of biofouling in ship hull turbulent boundary layers. *Trans Rina Part A*. 159(A3):291-300, 2017.

21. LIVINGSTON SMITH, S, BSRA resistance experiments on the LUCY ASHTON – Part IV, Miscellaneous investigations and general appraisal. *Trans Rina*. 97(4): 525-561, 1955

22. TANIGUCHI, K., FUJITA, T., Comparison of velocity distribution in the boundary layer between ship and model. 127:13-21

23. LEWTHWAITE. J.C, MOLLAND. A. F, THOMAS.K. W. An investigation into the variation of ship skin frictional resistance with fouling. *Trans RINA*. 127:269 – 284, 1984

24. POGGI, D., PORPORATO, A., RIDOLFI, L., An experimental contribution to near-wall measurements by means of a

special Doppler anemometry technique. *Exp in Fluids*. 32:366-375, 2002

25. HARDALUPAS, Y., TAYLOR, A. M. K. P. The identification of LDA seeding particles by the phase-Doppler technique. *Exp in Fluids*. 6(2):137-140-375, 2002

26. CLAUSER, F. H., 1954. Turbulent boundary layers in adverse pressure gradients. *J. Aeronaut Sci*. 21, 91–108.

27. CLAUSER, F. H., 1956. The turbulent boundary layers. *Adv. Appl. Mech*. 4,1–151.

28. MARUSIC, I., CHAUHAN, K. A., KULANDAIVELU, V., HUTCHINS, N., 2015. Evolution of zero-pressure-gradient boundary layers from different tripping conditions. *J. Fluid Mech*, 783, 379-411

29. CHAN, L., MACDONALD, M,. CHUNG, D., HUTCHINS, N., OOI, A., (215). A systematic investigation of roughness height and wavelength in turbulent pipe flow in the transitionally rough regime. *J. Fluid Mech.*, 771:743–777.

7. AUTHORS

BIOGRAPHY Dr Bagus

Nugroho
Dr Bagus Nugroho is a Postdoctoral Research Fellow in the Department of Engineering and Environment at The University of Southampton and visiting Research Fellow in the Department of Mechanical Engineering at The University of Melbourne. Dr Nugroho's research interests are in wall-bounded turbulent flow, passive flow control, drag reduction mechanism, and flow over rough surfaces. He also looks into environmental fluid flow, particularly at the effects of biofoulings on marine engineering applications.

Dr Rio Baidya
Dr Rio Baidya is a Postdoctoral Research Fellow in the Department of Mechanical Engineering at The University of Melbourne. His research interest is in experimental turbulent boundary layers, with focus on the effects of perturbations and roughness.

Muhammad Nizar Nurrohman Muhammad Nizar Nurrohman is a graduate student at the Dept of Naval Architecture, Institut Teknologi Sepuluh Nopember (ITS).

Adi Kurniawan Yusim
Adi Kurniawan Yusim is a lecturer in the Dept of Mechanical Engineering at Jember University. He graduated from, Institut Teknologi Sepuluh Nopember (ITS) in Naval Architecture. His field of study is in ship hydrodynamics.

Dr Fredhi Agung Prasetyo
Fredhi Agung Prasetyo holds the current position as Senior Manager of Research & Technical Application Division Research & Development, Biro Klasifikasi Indonesia. He has main research interest on fatigue & fracture, corrosion protection, analysis on ocean environmental effect to the maritime & ship safety and environmental protection.

Mohammad Yusuf
Mohammad Yusuf is the Director of Armada, Engineering, and Logistic for PT Dharma Lautan Utama. Mr Yusuf has an extensive 17 years of experience in engineering, operation, and maintenance for various types of passenger ship.

A/Prof I Ketut Suastika
Ketut Suastika is an Associate Professor in Hydrodynamics at the Department of Naval Architecture, Sepuluh Nopember Institute of Technology, Surabaya, Indonesia. A/Prof Suastika is a researcher in the field of wave-current interactions, structural responses due to waves and current, ship resistance and seakeeping performance. His major contributions have been in modeling of ringing of offshore structures, wave propagation on a counter current, including wave blocking, ship wakes and hydrofoil applications to ship- resistance reduction and seakeeping-performance improvement.

Prof I Ketut Aria Pria Utama
I Ketut Aria Pria Utama is a professor of ship hydrodynamics at the Institut Teknologi Sepuluh Nopember (ITS). He obtained his PhD in 1999 from the University of Southampton. He is a fellow member of the Royal Institution of Naval Architects (RINA) since 2006 and member of the Indonesian Academy of Sciences (AIPI) since 2015. His research interests include ship resistance and propulsion, seakeeping, computational fluid dynamics, renewable energy, and maritime engineering education.

A/Prof Jason Monty
Jason Monty is an ARC Future Fellow, Associate Professor, and the Head of Department in Mechanical Engineering at the University of Melbourne. A/Prof Monty is an experimental researcher in Fluid Mechanics and also the leader of the Michell Hydrodynamics Laboratory at Melbourne, which houses several new andunique flow facilities and state-of-the-art instrumentation. A/Prof Monty has a broad range of research interests from arterial flows to wall-turbulence to wave mechanics. His major contributions to science have been in the field of turbulent boundary layers, which has led to new experimental programs in turbulent air-sea interactions and in-situ measurements of the boundary layer on a ship.

Prof Nicholas Hutchins
Nick Hutchins has been an ARC Future Fellow and is a Professor in Mechanical Engineering at The University of Melbourne. Prof Hutchins is an experimental researcher in Fluid Mechanics, specialising in turbulent boundary layers, rough wall flows, flow control of turbulent boundary layers, drag reduction and advanced flow diagnostics. His major contributions have been in the classification of large-scale structures in turbulent boundary layers, and also in quantifying errors in fluid measurement techniques.

Prof Bharathram Ganapathisubramani
Bharathram Ganapathisubramani is a Professor of Experimental Fluid Mechanics in the Aerodynamics and Flight Mechanics Research Group at the University of Southampton. He was previously a Senior Lecturer at Southampton (2010-2012) and a Lecturer at Imperial College London (2007-2010). There are three major strands in BG's research and he has made several unique contributions in each of them. The first strand is aimed at physics and control of turbulent flows in aero-/hydro- dynamic applications. Examples include quantifying the effects of scale interactions in turbulent shear flows, isolating large-scale mechanisms responsible for skin- friction drag and scaling of large-scale energy containing motions in smooth-wall turbulent boundary layers, effects of geometric modifications of bluff bodies on dynamics of turbulence and established scaling laws and using multi-scale geometries for flow control. The second strand is focussed on fluid dynamics of biological/bio-inspired systems. Work in this area includes CT-scan based simulation of fluid flow through lymphatic node, aeromechanics of active and passive membrane wings, evolution of flight and swimming performance of marine reptiles. The final strand, which serves as the link between the other strands of my research, is on development of new experimental and data reduction methods including pressure determination using planar and volumetric velocimetry data, temperature measurements in high-speed flows and innovative ways of examining the data.

DEVELOPMENT OF RULES-REGULATIONS FOR SHIP INTENDED TO SAIL IN DOMESTIC INDONESIAN WATERWAYS: Effect of Storm model

F A Prasetyo, M A Kurniawan and **S Komariyah,** Biro Klasifikasi Indonesia, Indonesia

SUMMARY

Kurniawan et al (2016) propose the zonation system of Indonesian water from the basis of oceanographic history. This zonation system will be introduced and be proposed to replace the common service area that is used by Indonesian Administrator and their recognized organization. Prasetyo et al (2017) conducted comparison study of storm history in the two different area with the equator line basis position. The study was analyzed the storm in the sub tropical area and that of sub – tropical area. The sub-tropical area is represented from North Atlantic and North Pacific Ocean, where the tropical area is presented with Indonesian waterways and near territory waters. In this paper, effect of comparison study of storm in Indonesian water in term of proposed zonation system by Kurniawan et al, will be discussed. The possibility of strengthen of zonation system from storm study are examined.

1. INTRODUCTION

The safety and security shipping are mandatory aspect that should be fullfiled by all stakeholder maritim industries (shipping company, flag/port administrator, recognized organization. Indonesian Shipping act no. 17 2008 [1] mentioned that safety and security shipping consist of safety and security of shipping activities while in navigation voyage, port and include with environmental protection. Based on forementioned sentence, the ship seaworthiness becomes one aspect that suport the fullfillment of the safety and security shipping. The ship seaworthiness is dependable to the operating area, and the operating area is denoted with 'Service area" notation that is defined from Flag Administrator or recognized organization.

The Biro Klasifikasi Indonesia – BKI is Indonesian classification society that is one of Organization recognized by Indonesian goverment to conduct survey and statutoria certification on be half Indonesian goverment for Indonesian Flagged ships [2]. To support this recognization, BKI develop its own rules and regulation on basis of many aspects include with the ship operation area. Since that BKI's vessel have operating area in worldwide voyage (international trading) and/or Indonesian waterways voyage (Indonesian domestic trading).

The ship's operating area is reflected to the Service area notation from Class society that is recognized organization of FlagState Administrator. The example of service notation that it qoutes from BKI rules for classification and construction for seagoingships, Rules for Classification & Survey [3] is shown as Figure 13. The Service notation with "Unrestricted Service" means that ship will operated for all global oceans, seas and waters area. The restricted ocean service (**P**) is limited to ship having operating area only from range of 200 nautical miles from nearest port of refugess. 50 nautical miles is ranging area for coastal service (**L**) and the rest of shallow water service is area with calm seas harbours, bays and others typical areas. These service areas are based on North Atlantic ocean enviromental conditions [4, 5].

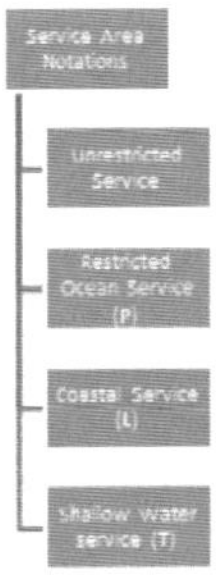

Figure 13, The Service area notation based on BKI Rules [3]

Since that the basis environmental of North Atlantic ocean might be extreme when used for the basis of Indonesia domestic trading, Kurniawan et al [6] compared three different zone in order to collect the actual condition of ocean environmental, and proposed the zonation system of service area dedicated for Indonesian waterways. Furthermore, Prasetyo et all [7] studied the applicability of storm model for the Indonesian tropical ocean. The historical wave data and storm data might be used as prime data for the development of service area that its dedicate to Indonesian water. In this paper, the effect of storm data in conjuction with wave height histories are analyzed and discussed.

The paper is organized as the following: Sect.2 presents the oceanographic data used in both study. Section 3 presents the analysis proses of oceanographic data,. Section 4 presents the obtaining results and discussion, and finally, the conclusion is given.

2. OCEANOGRAPHIC DATA

The development of service area definitioan is based on the actual condition of ocean. The actual condition of Ocean could be represented from the oceanograhic data collecting. The oceanograhic data has variation source, there are visual observation, instrumental observation, hind-cast data. The visual observation is performed from onboard or from coastal station inshore, while instrumental observation is provided from instrument measurement (buoy, wave radar, satellite). Thus, hind-cast data is a numerical model integration of both of observation data and instrumental data where no observation are recorded. The procedure to gather all observation data are under supervision of World Meteorological Organization (WMO) [8]. Recently, the oceanograhic data is mainly provided from the generation of the hind-cast model. Many study support the accuracy of hind-cast model, and that of collected from visual observation, instrument observation as well as buoy data or satellite data [9].

In this study, the oceanographic data that collected from hindcast data of European centre for medium-range weather forecasts (ECMWF)[10], is significant wave heigh, H_W; mean period, T_W and wave direction, θ. The data was collected for more than 30 years (1979 – 2015) within 4 times per days (06.00am, 12.00am, 18.00pm, 24.00pm). The data of Indonesian waterways is utilized from ranging area between 90°E to 140°E and 10°N to 10°S (see Figure 14).

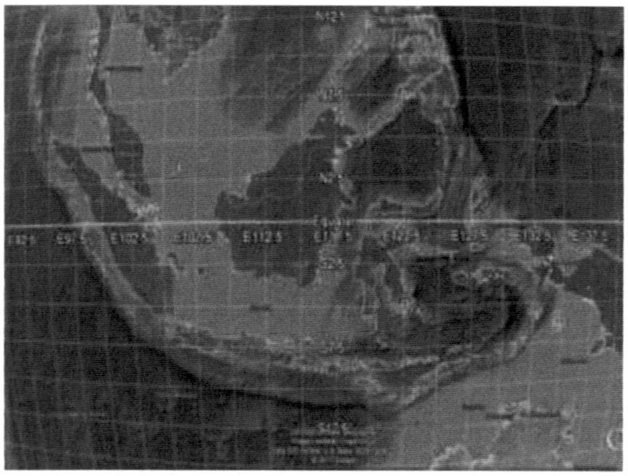

Figure 14 The map of area of Indonesian waters that is used for data collecting. [Source: Google Earth].

3. ANALYSIS OF OCEANOGRAPHIC DATA

3.1 (a) TIME HISTORY OF WAVE HEIGHT

Kurniawan et al [11] conducted study in order to develop the wave scatter map for Indonesian waterways (see Figure 15). Based on this map, the service area **P**, **L**(50) and **L**(20) are plotted. The H_W history of selected data points for each selected service area are analyzed based on 0.75degree and 1.0 degree lattice in order to simplify the numerical analysis. The total selected data location is ± 13800 points.

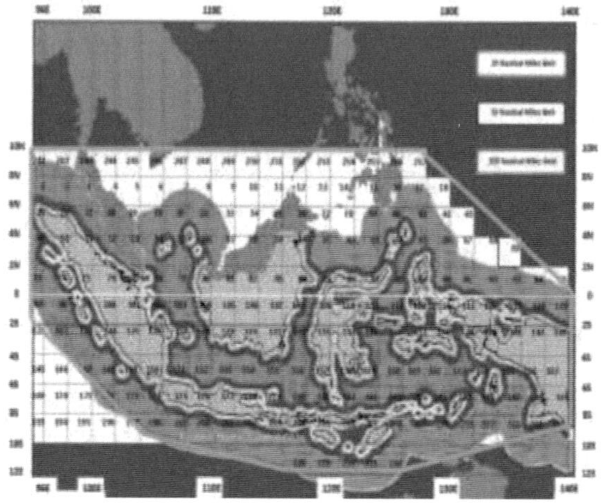

Figure 15 Wave scatter map of Indonesian waterways. The plotting of Service area based on Classification Society Rules[3].

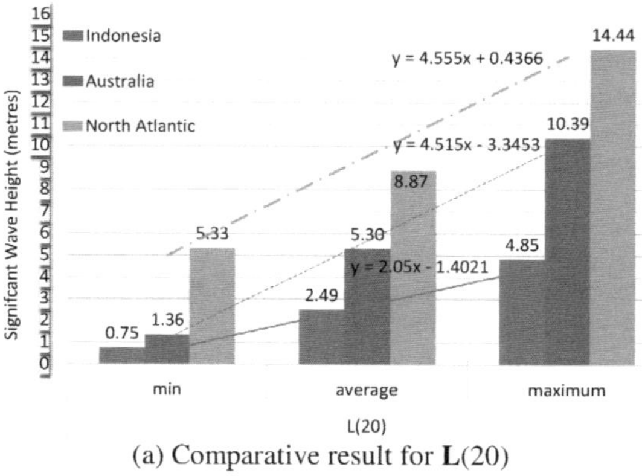

(a) Comparative result for **L**(20)

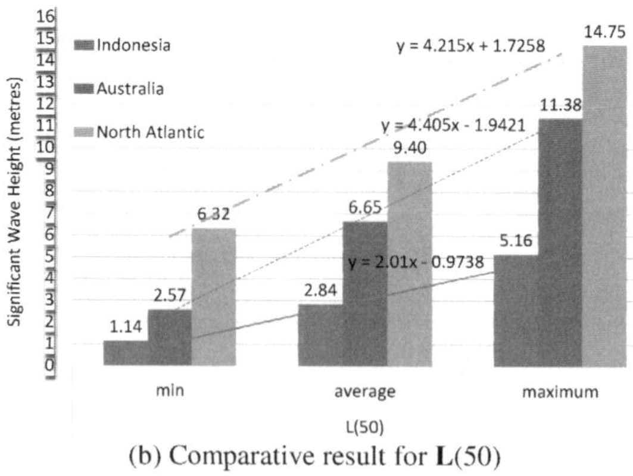

(b) Comparative result for **L**(50)

Figure 16 Comparison Three water area, North Atlantic Ocean, Australian waters,

Indonesian waters. (a) is the result for service area L with 20 nautical miles from coastal, and (b) is that of 50 nautical miles.

The selected locations on each service area of Indonesian water also compared by [6] with North Atlantic Ocean and Australian waters. Generally, it is reported that ratio between North Atlantic Ocean and Indonesian waters is 4. Thus, the ratio between Australian waters and Indonesian waters is 2.

In deeply discussion, as seen on Figure 16, the comparative of three different water area at two range of service area **L** are given. Two different coatal range of L are plotted, these are 20 nautical miles and 50 nautical miles. These figures show the comparative of trendline of minimum to maximum H_W include average of H_W for each water area. It is found that both of North Atlantic and Australian waters have the same gradients, that is about 4.2 to 4.5. The gradient of Indonesian waters is about a half of either of North Atlantic Ocean and Australian waters.

The above paragraph indicates that the application of service area by using the basis of North Atlantic ocean may not be suitable if they are applied at Indonesian waters.

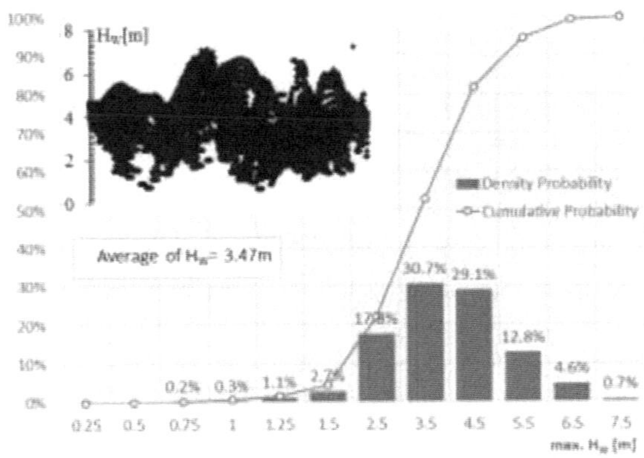

Figure 17 The histogram of H_W distribution on Indonesian waters [Figure 15]. The H_W distribution is also detailed.

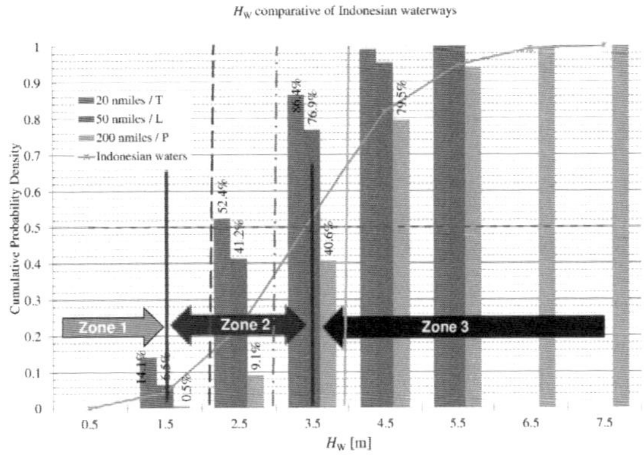

Figure 18 The comparative of H_W on service area (P, L, T) of Indonesian waters. The proposed zone condition also be drawn.

3.1(b) PROPOSED ZONATION AREA

Kurniawan et al [6] analyzed the time history of H_W at data locations with 43680 points. Figure 17 shows histogram of H_W distribution on Indonesian water, in which the range of H_W distribution also be plotted. The distribution of H_W is ranged from 0.68 meter to 7.23 meter, with average value is 3.47 meter. The distribution 's coeficient of variance is 0.34. The histogram shows 1meter discrete of H_W, exclude H_W which is lower than 1.5 meter, the 0.25meter discrete is applied. This figure shows that H_W ranges from 3.5 meter to 7.5 meter have population, 47.25%. The H_W ranges from 1.5 meter to 3.5 meter have popululation, 52.75%.

Furthermore, Figure 18 shows the comparative of H_W on Indonesian waters. The discrete 1meter H_W is also applied. The service area (**P, L, T**) on Indonesian waters are displayed. Based on the last report of Kurniawan [6], the average of H_W on the service area **P, L, T** are 3.74 meter, 2.84 meter and 2.49 meter. If it is compared with all Indonesian waters, 60% H_W in **P**, 23% H_W in **L** and 14 % H_W in **T** are located on cluster with H_W larger than 3.5meter. The ranged '1.5 meter $\leq H_W \leq$ 3.5 meter' consists of 35% H_W of **P**, 70.5% H_W of **L** and 71.9% of **T**. Thus, The ranged of "$H_W <$ 1.5 meter" consists of 0.5% of **P**, 6.5% of **L** and 14.1% of **T**.

It is concluded that the H_W could be clustered in three categories. The clustered system is defined on Table 1.

Table 1 The definition of zone clustering systems [6]

Zone	Definition
I	H_W < 1.5 meter
II	1.5 meter ≤ H_W ≤ 3.5 meter
III	3.5 meter < H_W

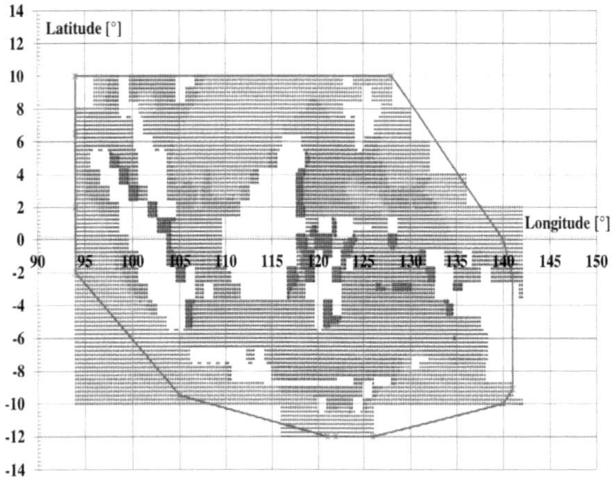

Figure 19 The mapping of point locations of Indonesian waters based on zone clustering as define in Table 1.

The combination of data locations of the source that is used in measurement of H_W and that of zonation area based on clustering system of Table 1 is plotted in Figure 19. This figure shows the graphical border and separation of zone clustering system in point view of maximum H_W around Indonesian waters. This figure describes zones separation with different colour. Zone I is by green colour, Zone II is by blue colour and Zone III is by red colour.

3.2 STORM MODEL ANALYSIS

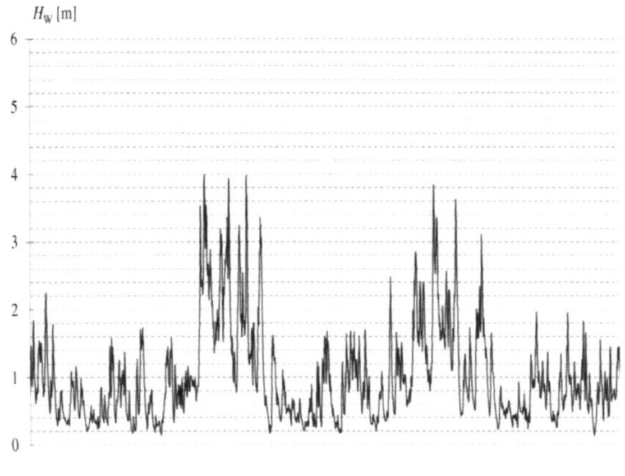

Figure 20 Typical time histories of H_W, which is used to generated storm model.

In the study of the extreme condition of ocean in tropical/ equator region, Prasetyo et al [7] developed the analyses system based on storm model. The Indonesian waters is choosed as representation of the tropical/equator region. The storm profil is generated from time history of H_W in the selected locations as described in previous paragraph. The example of typical time history of H_W is shown as Figure 20.

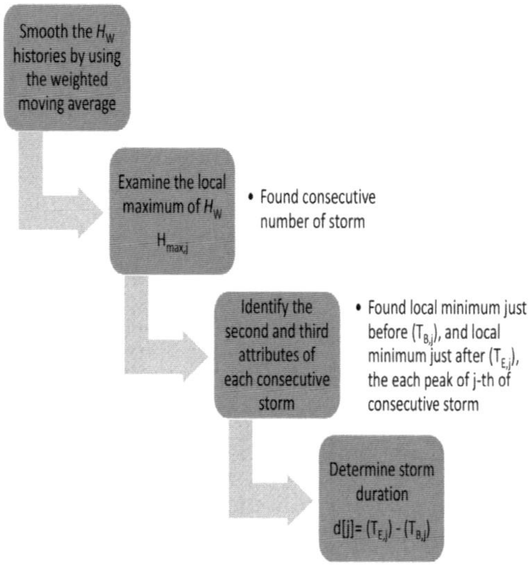

Figure 21 Storm profile analzing procedure

The storm profile is examined from the changing nature of H_W time histories by using specific procedure [12]. The procedure could be drawn as Figure 21. The storm profile will be identified based on storm class. Storm class is maximum local H_W ($H_{max,j}$), that is identified during storm generation as described in Figure 21. The storm class is considered with integer number and discrete it every 1 meter.

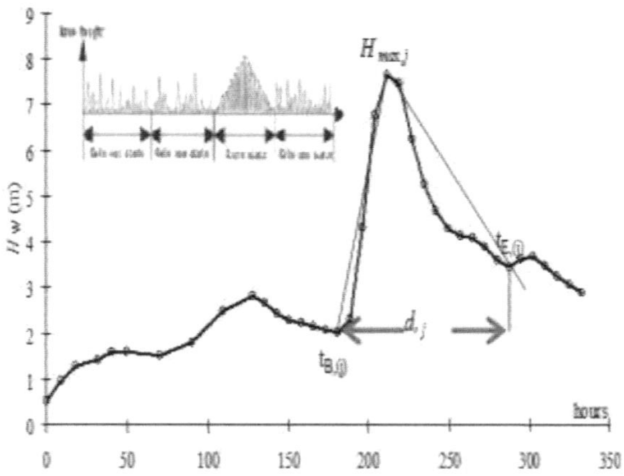

Figure 22 The storm definition based on Equivalent triangular storm [12].

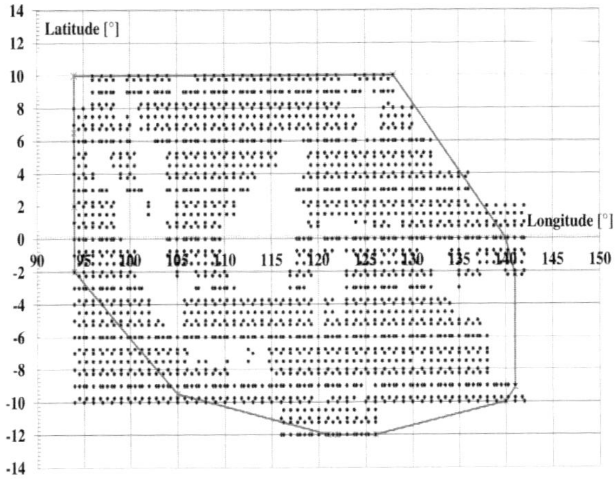

Figure 23 Point location that the storm is identified at the chosen location of Indonesian waters.

The results, Prasetyo [43] reported that the storm is not identified at all point locations (see Figure 23). The number of storm class occurs in 1936 points from 43680 points. The identified storm/ extreme wave condition on Indonesian waters is ilustrated in Figure 24. The average duration of storm in Indonesian waterways ranges from 2.0days to 2.857days, with the maximum duration is ranging from 3days to 8.75days and minimum duration is about 1.25days. The storm duration fluctuation could be regressed by using the linear interpolation of lognormal of exceedance probability, as present in Figure 25.

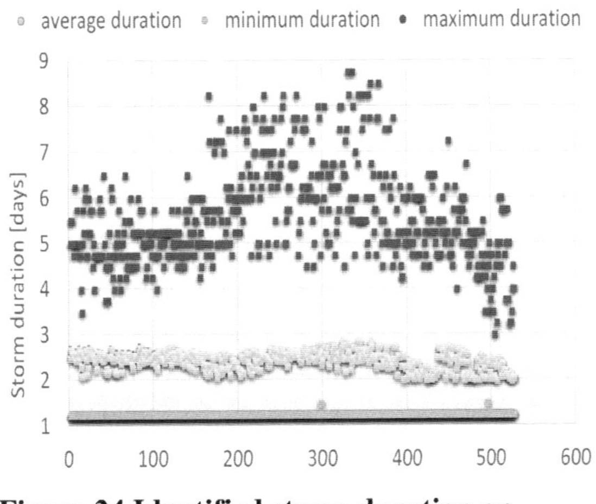

Figure 24 Identified storm duration on Indonesian waters

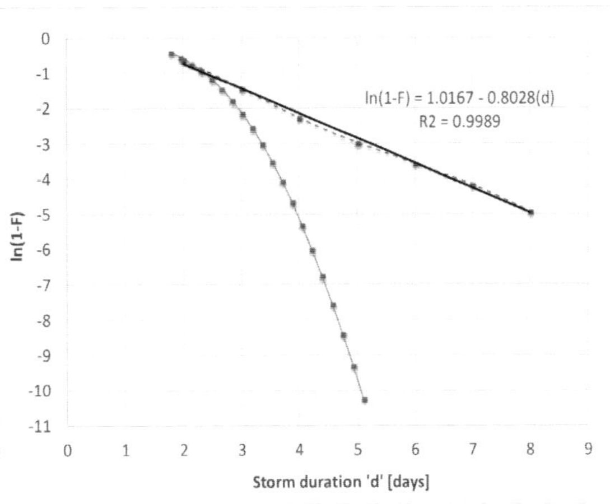

Figure 25 The regressed storm duration on Indonesian waters

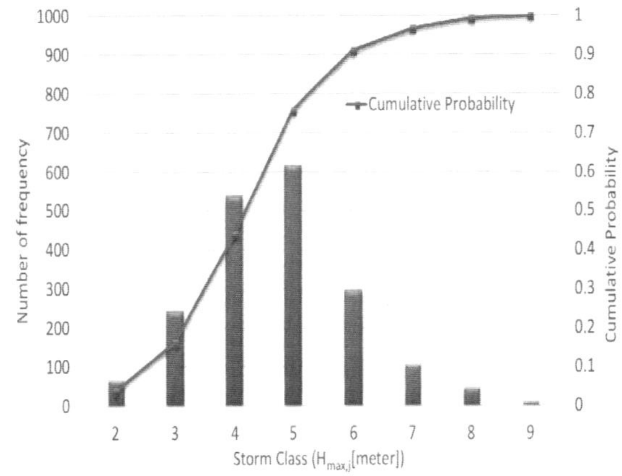

Figure 26 Histogram of maximum storm class from identified points locations along Indonesian waters. The cumulative probability distribution is plotted.

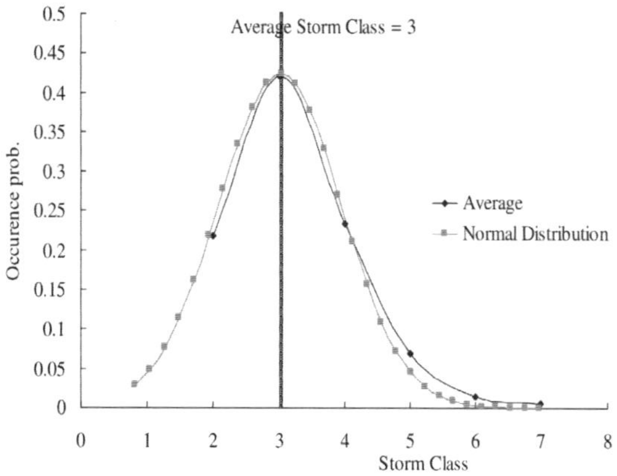

Figure 27 The regressed storm classes on Indonesian waters

It also be reported that the storm is profiling from cresecendo de crescendo of storm class, while the histogram of storm class are figured in Figure 26. The minimum of storm class is 2 meter and maximum of storm class is 9 meter. Furthermore, the statistic characteristic or value of storm class that identified on Indonesian water could be distributed by using Normal Distribution. Figure 27 shows the regressed Normal distribution and that of occurence probability for all storm classes, and it is fitted in good agreement.

4. DISCUSSION

From above paragraph, the maximum H_W and storm class for each data locations could be generated. The wave height histories are clustering based on Table 1. Since that the storm profile is consist of crescendo de crescendo of wave height (see Figure 28), it is also noted that storm class could also be clustered as well as the wave height histories.

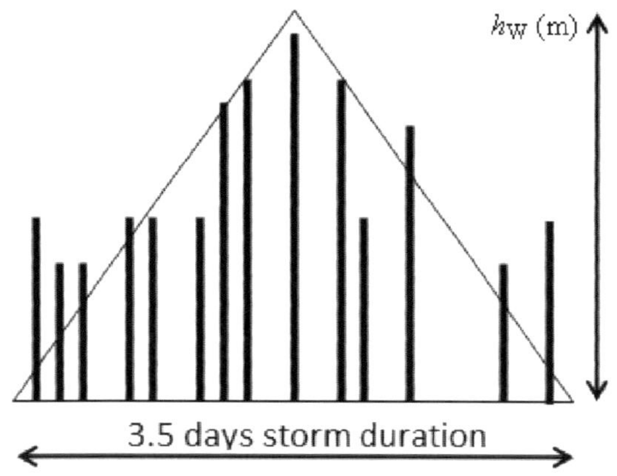

Figure 28 Typical storm profile

Figure 26 indicates that the majority distribution of storm class occurs while it is larger and equal than 4 meter storm class. It is denoted with Class A. The other storm class group is distributed when storm class is ranging from 2 meter to 4 meter. This is denoted with Class B. The all groups of storm class location are shown in Table 2.

Table 2 Storm class group point location

Notes	Points	Longitude	Latitude
Class A	9216	94°E	10°S
	10240	96°E	4°N
	10974	97.5°E	6°N
. . .			
Class B	10239	96°E	3.75°N
	10240	96°E	4°N
	10953	97.5°E	0.75°N
. . .			

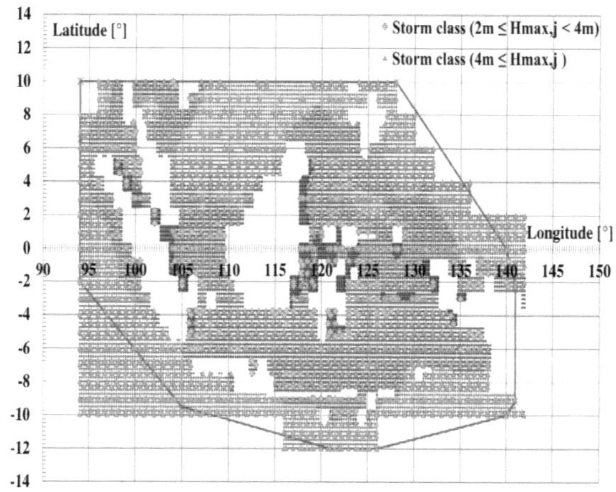

Figure 29 Plotting of storm classes for each longitude and lattidue locations.

Table 3 Correlation between H_W histories and storm class cluster.

Cluster	H_W histories	Storm Class ($H_{max,j}$)
I-C	$H_W < 1.5$[m]	---------
II-B	1.5[m] $\leq H_W \leq$ 3.5[m]	2[m] $\leq H_{max,j} <$ 4[m]
III-A	3.5[m] $< H_W$[m]	4[m] $\leq H_{max,j}$

Based on storm class grouping and its locations, all storm class locations are plotted in Figure 29. Two group of storm classes are plotted, that is group A and group B, as described in previous paragraph. The maximum wave height for each point locations also be plotted in this figures. It is shown that storm classes with group B (2m ≤ $H_{max,j}$ < 4m) is identified in the same locations with points of zone I and II of Table 1. On the other group, group A of storm Classes, its identify with that of points of zone III of Table 1. It is concluded that the cluster of storm classes be identified in Indonesian waters may be clustered as the same as wave height histories [Table 1]. The clustering system of both of wave height histories and storm class is described on Table 3. The illustration of joint cluster between wave height history and storm class is shown as Figure 30. This figure shows that the joint cluster is in line from the cluster II – B and III – A. This means that two clusters are medium and high severe area. The calm area become the cluster I- C.

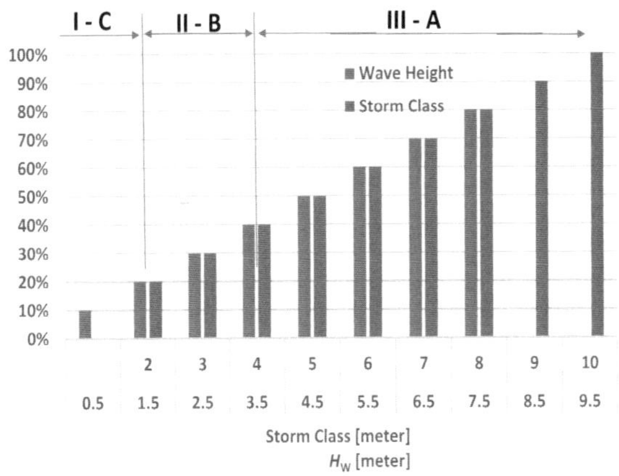

Figure 30 Ilustration of the correlation between H_W and Storm class cluster.

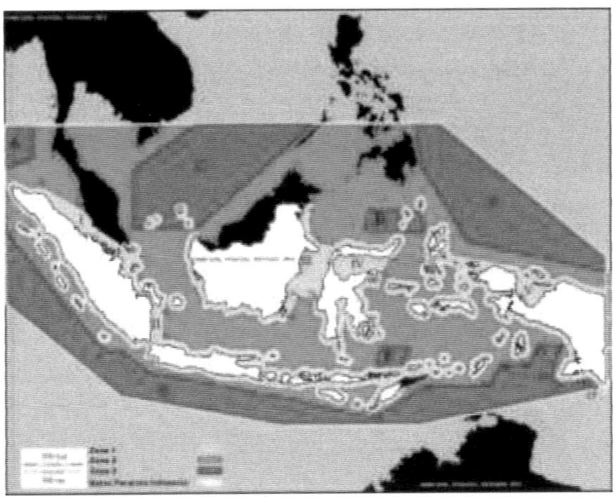

Figure 31 The proposed service area of Indonesian waterways, based on zonation system [6].

In connection with, wave height histoy, storm class and service area on the zonation basis, it could be concluded as following:

a. Time histories of H_W at Indonesia waterways are examined, and its compare with that of North Atlantic Ocean and Australian waters.
b. The cluster of H_W at Indonesian waterways is proposed and it consists 3 cluster based on H_W:
   i. H_W < 1.5m
   ii. 1.5m ≤ H_W ≤ 3.5m
   iii. 3.5m < H_W

c. The storm/extreme wave occurs in Indonesian waters is identified by using storm model. The storm model has each storm profil that identify by using storm class. The cluster of storm is divided based on two conditions, that is:
   i. $H_{max,j}$ < 4m
   ii. 4m ≤ $H_{max,j}$

d. The characteristics of Indonesian waters on the wave height histories, storm duration and storm class have widely range of scatter.

In combination with ship service, distance from coastal/ range of refuge, the Indonesian water enviromental and its analysis, it is concluded that service area based on zonation system could be applied on Indonesian waters as be shown in Figure 31.

5. CONCLUSIONS

The study on rules regulation development intended to ship sailed on Indonesian waters has been established. The continuity of work is started from wave scatter map analysis, comparative study on the three different waters; North Atlantic Ocean, Australian water and Indonesian waters. Thus, it is continued by work on extreme wave condition in Indonesia water as a part of equator/ tropical area.

The current work gives the joint cluster group for both of wave height histories and storm class, as representation of analysis on Indonesian waters as well as the extreme wave condition, as following :

I-C	H_W < 1.5[m]	
II-B	1.5[m] ≤ H_W ≤ 3.5[m]	2[m] ≤ $H_{max,j}$ < 4[m]
III-A	3.5[m] < H_W[m]	4[m] ≤ $H_{max,j}$

Finally, the service area based on zonation system is proposed to be implemented for Indonesian domestic voyage.

8. ACKNOWLEDGEMENTS

Met-ocean data from ECMWF is owned by the European Centre for Medium-Range Weather Forecasts (ECMWF) and accessed and downloaded from http://data-portal.ecmwf.int/data/d/interim_full_daily/.

This study is part of research activities of the "Wave-structure-material" Research and Development Division conducted by Fredhi Agung Prasetyo, Mohammad Arif Kurniawan and Siti Komariyah. The authors would like to thank you for all assistance given in this research activity.

9. REFERENCES

1. Undang-Undang Republik Indonesia, Nomor 17 Tahun 2008, tentang Pelayaran, 2008.

2. Direktorat Jenderal Perhubungan Laut, DJPL, Surat Keputusan Direktur Jenderal Perhubungan Laut Nomor PK.204/1/3/DJPL-16 tentang Penunjukan Biro Klasifikasi Indonesia (Persero) sebagai Recognized Organization untuk melaksanakan survey dan sertifikasi Statutoria atas nama Pemerintah pada kapal Berbendera Indonesia, 2016.

3. BKI, "Rules for Classification and Surveys, Volume I", Rules for the classification and construction of seagoing steels ships part 1, BKI, 2016.

4. IACS, "IACS Reccomendation No. 34 Standard Wave Data", IACS, 2001.

5. IMO MSC resolusi MSC.287(87), "Adoption of the International Goal-Based Ship Construction Standards for Bulk Carriers and oil Tankers", IMO, 2010.

6. Kurniawan M A, Prasetyo F A, Komariyah S, "A comparison of three different water areas and its influence for development of rules regulations", proceeding of Asian-Pacific Technical Exchange and Advisory Meeting (TEAM) 2016.

7. Prasetyo FA, Kurniawan MA, Komariyah S, Rudiyanto, Herawan T: Storm application at Indonesian Tropical Ocean, ICCSA 2017, part IV, LNCS 10407, pp.732-745, Springer International Publishing AG, 2017

8. World Meteorological Organization: 'Guide to wave analysis and forecasting', World Meteorological Organization (WMO) , 1998.

9. Mao W, Prasetyo FA, Ringsberg JW, Osawa N: ' A comparison of two wave models and their influence on fatigue damage in ship structure, proceeding of 32nd the International Conference on Ocean, Offshore and Arctic Engineering (OMAE), ASME, 2013.

10. European Centre for Medium-Range Weather Forecasts (ECMWF), www.ecmwf.int.

11. Kurniawan M A, Prasetyo F A, Komariyah S, "Study on wave scatter mapping of Indonesia waterways based on hind-cast data", proceeding of Asian-Pacific Technical Exchange and Advisory Meeting (TEAM) 2014, ITU published.

12. Prasetyo F A, Study on advanced storm model for fatigue assessment of ship structural member, Doctoral dissertation, Osaka University Suita Japan, 2013

10. AUTHORS BIOGRAPHY

[**Fredhi Agung Prasetyo**] holds the current position of Senior Manager of Dept. Research and Technical Application, Research & Development Division. Beside managerial position, he has responsible for Wave, Structure and Material Research.

[**Siti Komariyah**] holds the current position of Senior Researcher at Dept. Research and Technical Application, Research & Development Division. She has responsible for Structure and Material Research.

[**Mohammad Arif Kurniawan**] holds the current position of Senior Researcher at Dept. Research and Technical Application, Research & Development Division. He has responsible for Wave and Structure Research.

PRELIMINARY DESIGN OF MOBILE FISH PLATFORM FOR LOCAL FISHERMEN

W. R. Hetharia, **F. Gaspersz** and **R. Latuhihin**, Faculty of Engineering - Pattimura University - Ambon, Indonesia

SUMMARY

Indonesian waters have a great potential of small pelagic and reef fishes. Those fishes are caught by local fishermen with traditional fishing devices. One of traditional devices used up to now is fish platform or '*bagan*' in local name. Problems in operation of *bagan* operated by local fishermen in Indonesia waters are solved by application of mobile platform. This fish platform is equipped with self outboard engines and multi-fishing gears. This design innovation ends-up with a mobile fish platform that can operate at any fishing grounds, several catching operations and some kinds of fishes. Initial surveys were conducted to find the proper fishing grounds and input design parameters. This mobile fish platform have been designed and construction phase is undergoing. The performance of mobile platform will be evaluated at sea trial in the future. The results of sea trial will be used for new design and application for local fishermen business groups.

Key words: design, mobile platform, mechanical system

NOMENCLATURE

[Symbol]	[Definition] [(unit)]
Δ	Weight displacement (tonne)
L_{OA}	Length overall (m)
L_{WL}	Length of waterline (m)
B	Beam (m)
B_{hull}	Beam of one hull (m)
T	Draft (m)
H	Deck height (m)
C_P	Prismatic coeficient
C_B	Block coeficient
C_M	Midship coeficient
LCB	Longitudinal centre of buoyancy (m)
KB	Vertical centre of buoyancy (m)
KG	Vertical centre of weight (m)
BM_T	Transverse radius of metacentre (m)
BM_L	Longitudinal radius of metacentre m)
GM_T	Transverse metacentre height (m)
GM_L	Longitudinal metacentre height (m)

1. INTRODUCTION

Indonesia is one of the big maritime countries in the world which have a great potential of fish resources. Any small fish business groups until international groups are involved in managing fish industry in Indonesia. Exploitation of fish resources are executed by using any kinds of fishing vessels and fishing gears. Coastal fishermen play the role also in exploitation the fishing resources by using any types of traditional fishing devices in form of personal or small groups. These situations cause their fishing business are not well develop. One of the traditional device which is still used by coastal fishermen until now is fish platform or '*bagan*' in local name.

The operation of traditional fish platform by coastal fishermen were followed by any problems such as: since fish platform is fixed at the anchor then it depends on how much fishes (small pelagic) arrive, the application of one fishing gear (lifting net), the effects of climate and sea current, line of fish migration, lighting system and short material construction lifetime. Those issues end-up with small catching product or economically less profitable.

Those problems in operation of traditional fish platform are solved by application of mobile fish platform. This mobile fish platform can be operated at any fishing grounds with 3 kinds of fishing gears (lifting net, bottom vertical line, trap net) for caching small pelagic fishes, coral fishes and any kinds of lobsters. Also, catching operation may be executed several times at night and day time. Mechanically, those three fishing gears make easy catching operation therefore, the catching product will increase.

Innovation of mobile fish platform were executed by a design process to obtain proper dimensions and configurations to satisfy target operation of fish platform. Design process of mobile fish platform principally similar to ship design process. The design process includes specifying input design, computation of design parameters, evaluation of design parameters and determination of output design. Particular study were executed in design process of multi fishing gears to ensure smooth catching operation. Blue print of output design was used to construct a prototype mobile fish platform and fishing devices.

2. LITERATURE REVIEW

2.1 FISH PLATFORM (*BAGAN*)

Technology of catching fish used by coastal fishermen in Indonesian waters generally are in traditional way. This issue results in less catching product and income. Many group of fishermen have developed their modern fishing devices however, many fishermen groups still use some traditional devices. One of the fishing devices still used until now is fish platform or *bagan* in local name as Figure 1. This fishing device use one fishing gear which is lifting net and operate at night time. Kind of fish cached in this device is small pelagic (local and Latin name) such as *"layang"* (*decapterus spp*), *"kembung"* (*Restrelinger*), *"layur"* (*trichiurus lepturus*), etc. This mobile fish platform has two types which are raft and boat platforms. Raft platform means two hulls of platform are made of bamboo raft while boat platform means two hull of platform are made of small wooden boat (Sudirman, et. Al 2004; Sudirman, et. Al 2011; Usemahu, et. Al 2004; Tadjjudah, 2009). Fish platform consists of two left and right hulls serve as buoyant. Those two hulls are connected with wooden beam or bamboo. Fish platform is fixed at anchor to prevent if from drifting at sea.

Beside bagan, there is another fishing devices named '*rumpon*'as shown in Figure 2. *Rumpon* has a function of collecting fishes and furthermore catching operation will executed by purse seiner and designed mobile fish platform. Some researches executing by the author in designing and constructing Rumpon may be applied in constructing Rumpon in the future (Hetharia, 2000; Hetharia, 2001; Hetharia 2002).

Catching operation of fish platform is executed in the night time with support of light. The brighter the light the more chance to obtain many fishes. This is because fish behaves as positive phototoxic which means tendency of close the light (Ardi, 2009; Suherman, 2005). The role of light is to collect the fish in a "catchable area" and the catching process will be executed by net or other catching devices (Sudirman dan Malawa, 2004). Other requirements in operation fish platform are the condition of waters close the coast or in the bay protected from excessive wind, current and waves and the water deep less than 50 m as well as the water is in clear condition (Usemahu, et al, 2004).

Figure 1. Traditional Fish Platform (Type 2 and 3 Hull)

Figure 2. Types of *Rumpon*

Other factor which affect the success of catching operation of fish platform is determination of proper fishing ground and water depth (Kusuma dkk, 2014; Purbayanto, 2014). Fish platform is required to add fish aggregating device in order to attract many fishes to stay under the platform (Sudirman dan Malawa, 2004). Baskoro et. Al (2006) suggested the addition of Rumpon as device to collect many fishes under fish platform to increase catching product.

Solution offered in this work includes:

- Mobile fish Platform is equipped with self propulsion device (outboard engines) in order to move to other fishing grounds for catching operation.
- Mobile fish platform is equipped with multi fishing gears (Figure 3) in order to catch other kind of fishes such as grouping *grouper* (*grouper*), snapper (*snaper)* and any kinds of lobster.
- Mobile fish platform is equipped with other *rumpons* therefore the catching operation may be executed several times a night.

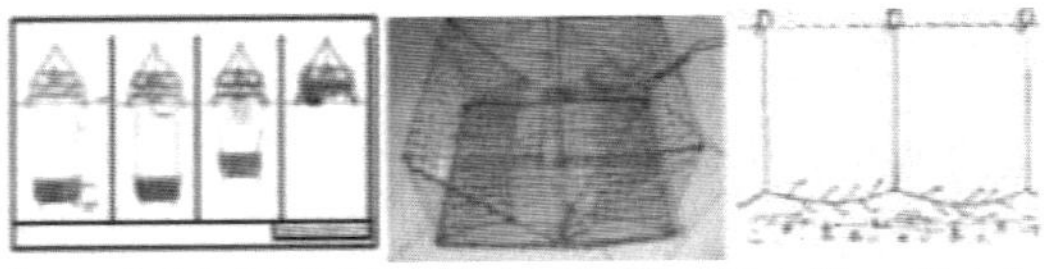

Figure 3. Multi Fishing Gears (Lifting Net, Trap Net dan Bottom Long Line)

2.2 DESIGN PROCESS OF FISH PLATFORM

Design process of mobile fish platform principally similar to that of seagoing vessels. A main difference lays on lay-out, function and outfit. Principally, design procedure similar to design procedure of ship which is an itteration process in early design phase. The output of first itteration should be evaluated, analyzed and modified untuil design process satisfy the design objective and requirements where this process is expressed in spiral design. A number of itterations are needed to satisfy a proper design result (Gale, 2003; Watson, 1998; Lamb, 1969). The design process need input design, design parameters, computatoins, determination of design output and design specification. Design parameters of mobile fish platform include: mission reruirements, dimensions and hull geometrical forms, space lay-out, weight components and their centres, trim and stability, speed and engine power, stucture components, prime mover and systim, ship outfit, fishing gears and cost.

Design process of mobile fish platform begins with mission requirements. Design parameters were computed and analyzed at each itteration in a design spiral until determination of proper parameters of mobile fish platform. A design consideration was taken into account that, at this moment, design parameters have not stated as 'optimal' for the reason that design mobile platform is a prototype. It is needed more improvements and corrections during the real operation of platform at certain operation period.

A main requirement in design mobile platform is a big amout of cathing product. This requirement may be obtained when the fishes are exists under the platform, placement of mobile platform at proper fishing grounds, a good lightning system, a wide net area and high speed of net hauling. A requirement of many fishes exist under the platform is obtained when operating the mobile platform, meanwhile a good lightning system dan wider net area are determined during the design process. However, the requirement of wider net area affect other design parameters.

Input design parameters are fixed as follows: net area $A_{net} = 67\ m^2$, cathing capacity 2.5 tons, service speed $V_{platform}$ = 6 knots. Other design considerations are platform profile is catamaran type, hull material is Fibreglass Reinforced Plastic (FRP), vertical structure of hull is Class I woods, transverse structure is steel profiles combined with vertical Class II woods and one small control house above the transverse structure. The lay-out of platform includes: fish net is between two hulls, fish holds and other storages are in the main hulls, two units of outboard engines 2 x 45 HP are placed at transom stern, hauling winch is placed at the midle of right hull while electric generator is placed at the midle of left hull and control navigation and operation control as well as crew accomodation are placed at upper house.

Design process started with a sketch and determination of dimensions, hull forms and lay-out of the platform. Weight components and their centres were computed in details. Trim and stability were determined based on weight details. Transverse structures were computed to ensure the strenght of mobile platform. Type and capacity of propulsion system were determined at design phase while service speed will be find during sea trial. Ship outfits and cathing devices were fixed during design process. Some references concerning previous design and operation of small fishing boats material FRP from the author were aplied in this work (Hetharia, 2000; Hetharia, 2002; Hetharia 2003).

3. METHODOLOGY

This research ends-up with a prototype of mobile fish platform and mechanical system of multi fishing gears. This output are obtained by the following steps:

- Survey of fishing ground location of mobile platform: the objective of this work is to obtain the dimensions and configurations of traditional fish platform and *rumpon*, potential fishing grounds and other problems experienced by local fishermen
- Design of mobile fish platform and mechanical system of fishing devices: the objective this work is to obtain a proper fish platform and mechanical system of catching devices. The output of this step was blue print and working drawings to be used in constructing mobile platform and mechanical system of catching devices.
- Constructing of mobile platform and mechanical system of catching devices : the objective of this work is to obtain a proper mobile fish platform and mechanical system of catching devices.

This research will be continued in the future for sea trial of mobile platform and catching devices. This work is to confirm the design output, evaluation and improvement of technical aspect and economical feasibility as a reference for new design and local fishermen groups.

4. RESULT AND DISCUSSION

4.1. RESULTS OF RESEARCH

Research activity was started with a survey of location of fishing grounds for mobile fish platform. The location of fishing grounds was fixed around Ambon Island with maximum of 3 milles from the coastline (Figure 4).

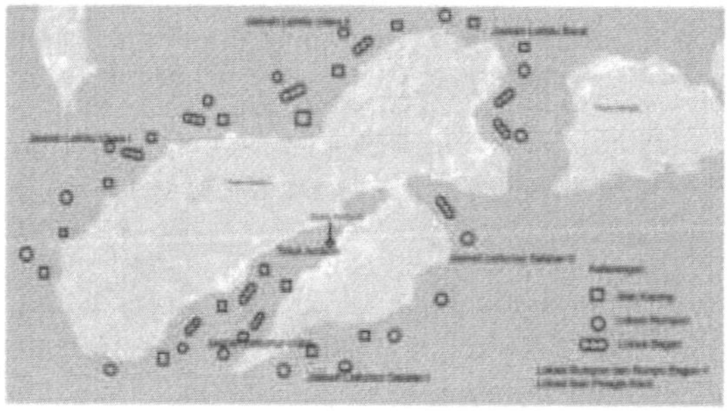

Figure 4. Location of Fishing Ground Mobile Platform

Results of design dimensions and hydrostatic parameters are presented at Table 1. Generall arrangement and body plan are showed in Figure 5.

Table 1. Main Dimensions and Hydrostatic Parameters of Mobile Fish Platform

No	Parameter	Symbol	Value	Unit
1	Length overll	L_{OA}	10,10	m
2	Length of waterline	L_{WL}	9,91	m
3	Beam	B	8,00	m
4	Beam of one hull	B_{hull}	0,90	m
5	Draft	T	0,74	m
6	Deck height	H	1,00	m
7	Displasement	Δ	9,40	ton
8	Prismatic coeficient	C_P	0,86	
9	Block coeficient	C_B	0,79	
10	Midship coecisient	C_M	0,92	
11	Waterline coeficient	C_{WP}	0,94	
12	Longitudinal centre of buoyancy, from AP	LCB	4,66	m
13	Longitudinal centre of floatation, from AP	LCB	4,71	m
14	Vertical centre of buoyancy, from BL	KB	0,412	m
15	Vertical centre of gravity, from BL (computation)	KG	2,960	m
16	Transverse Radius of metacenter	BMt	21,05	m
17	Lognitudinal Radius of metacenter	BMl	11,94	m
18	Transverse metacentric	GMt	18,50	m
19	Longitudinal metacentric	GMl	9,40	m
20	Immersion (TPC)	TPC	0,152	Ton/cm
21	Righting moment 1 deg = GMt.Disp.sin (1)	RM	2,96	Ton.m

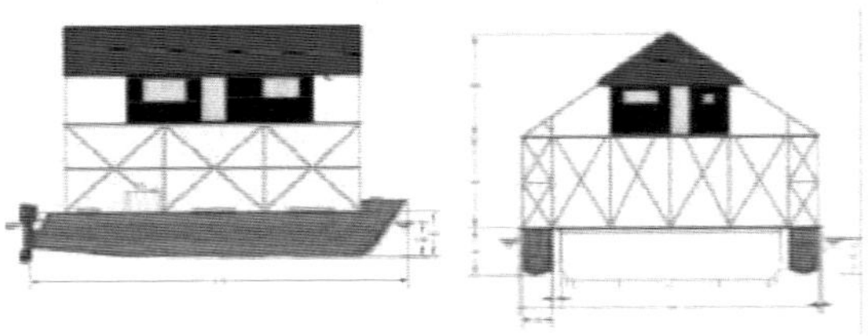

Profile and front view

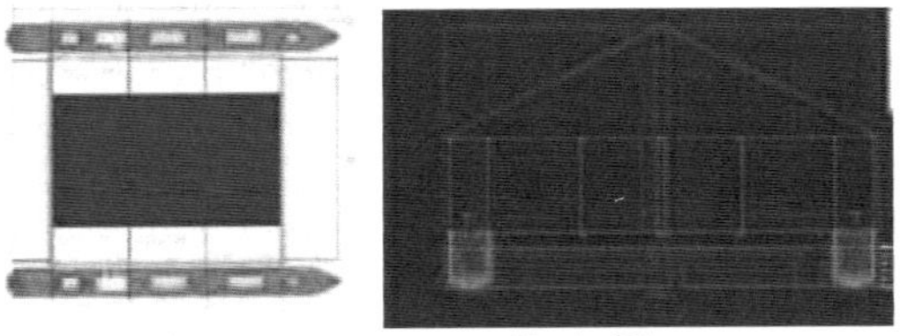

Top view Body Plan

Figure 5. Generall Arrangement and Body Plan Mobile Fish Platform

Design Spesification:
- Net area: 67.20 m^2
- Deadweight: 4.25 tonnes
- Lightweight: 5.15 tonnes
- Payload: 3.45 tonnes
- Displacement: 9.40 tonnes
- Electric power: 5 kW
- Crew : 4 persons
- Winch power: 5 kW
- Prime movers: 2 x 40 hp (outboard engines)
- Fishing ground: Around Ambon Island

Spesification of Catching Devices:
- Lifting net:
  - Area : 67.20 m^2
  - Depth : ± 20 m
  - Light : 0.5 kW,
  - Light control: electric dimmer,
- Vertikal line:
  - Depth: ± 100 m
  - Number: 20 units
  - Number of hook per unit : 5
- Trap net:
  - Type: folding net
  - Depth: ± 50 m

4.2. CONSTRUCTION OF MOBILE PLATFORM

The steps of construction of mobile fish platform are: construction of template framework, hull template and hull. Lines plan of hull were transferred into mould loft (scale 1:1) to draw body plan, profile plan and half-breadth plan. Next step was cutting off frames at plywood sheet, arrangement of frames in longitudinal direction and assembly the template framework and setting framework shell. Covered the template framework with material Fibreglass Reinforced Plastic and wooden ribs to create hull template. Two hulls of mobile platform were developed up to this time and the work is continuing to assembly the whole construction of mobile fish platform.

Figure 6. Construction of Frame Template

Figure 7. Contruction of Hull Template

Figure 8. Construction of Hull

4.3. DISCUSSION

The success of operation of mobile fish platform highly depend on its placement at potential fishing ground and supported by technical aspects provided at mobile platform. The output of design were platform dimensions, design specification and catching device have been obtained in this design process. Determination of parameters in this preliminary design stage include: net area, volume of some spaces at mobile platform, correlation between displacement and total weight as well as stability (Gale, 2003). Stability parameter was determined due to criterion of GM_0 > 0.15 m based on IMO Code on Intact Stability (Biran, 2003). The criteria will be a analyzed further in the future for large angle of inclinations. The speed and engine power will be found during sea trial test. The construction works is continuing and will be completed in the near future.

This mobile fish platform is a prototype of a new kind of fish platform. In prelimary design phase, there were no database as comparisson that could help to make easy computation of design parameters. Therefore in sea trial process at sea some input data and records will be required as reference for new future design.

5. CONCLUSIONS

Problems concerning operation of traditional fish platform were solved with application of mobile fish platform equipped with mechanical system of catching devices. Research works are executed for the construction of mobile platfom. Some conclusions were concluded such as:

- Research works of mobile fish platform were executed to fit the scenario which are early survey of fishing ground, design process and construction works.
- Some parameters of mobile platform were not determined at preliminary design stage, however those parameters will be obtained during the sea trial in the future in order to have improvement for the prototype platform and new design as well as refernce for local fishermen.

6. ACKNOWLEDGEMENTS

Special thanks is appreciated to The Direktur Jenderal Penguatan Riset dan Pengembangan Kementerian Riset, Teknologi dan Pendidikan Tinggi Republik Indonesia for its research grant "Penelitian Produk Terapan", 2017 s/d 2019.

7. REFERENCES

1. ARDI, S. A. H., 'Pola Persebaran dan Hasil Tangkapan Bagan di Teluk Pelabuhan Ratu', *Skripsi, FMIPA, Universitas Indonesia, Depok*, 2009.
2. BASKORO, M. S., Telussa R. F. dan Purwangka F., 'Efektifitas Bagan Motor di Perairan Waai, Pulau Ambon', *Prosiding Seminar Nasional Perikanan Tangkap -Menuju Paradigma Teknologi Perikanan Tangkap yang Bertanggungjawab Dalam Mendukung Revitalisasi Perikanan, Institut Pertanian Bogor*, 2006.
3. BIRAN, A. B., 'Ship Hydrostatics and Stability', *Butterworth – Heinemann, Oxford*, 2003
4. GALE, P. A., 'The Ship Design Process', *Ship Design and Construction – Chapter 5, Vol. 2, SNAME Publication, Jersey City, NJ, USA*, 2003
5. HETHARIA, W. R., 'Desain Rumpon Semi Terapung Untuk Beroperasi di Perairan Pulau Ambon', *Laporan Akhir Penelitian Dosen Muda, Direktorat Pembinaan Penelitian dan Pengabdian Masyarakat, Dirjen DIKTI, DEPDIKNAS*, 2000
6. HETHARIA, W. R., 'Desain Alat Tangkap Tuna Long Line yang Beroperasi di Perairan Pulau Ambon', *Laporan Akhir Program VUCER, Direktorat Pembinaan Penelitian dan Pengabdian Masyarakat, Dirjen DIKTI, DEPDIKNAS*, 2000
7. HETHARIA, W. R., 'Desain Lampu DC Otomatis untu Perlengkapan Rumpon yang Beroperasi di Perairan Pulau Ambon', *Laporan Akhir Program Vucer, Direktorat Pembinaan Penelitian dan Pengabdian Masyarakat, Dirjen DIKTI, DEPDIKNAS*, 2001
8. HETHARIA, W. R., 'Desain Sistim Jangkar Alat Pengumpul Ikan (Rumpon) yang Beroperasi di Perairan Maluku Tengah', *Laporan Akhir Program Vucer, Direktorat Pembinaan Penelitian dan Pengabdian Masyarakat, Dirjen DIKTI, DEPDIKNAS*, 2001
9. HETHARIA, W. R., 'Desain Kapal Penangkap Ikan Tuna Long Line 1,5 GT Material Fibreglass Reinforced Plastic (FRP) Untuk Beroperasi di Perairan Pulau Ambon', *Laporan Akhir Program VUCER, Direktorat Pembinaan Penelitian dan Pengabdian Masyarakat, Dirjen DIKTI, DEPDIKNAS*, 2002.
10. HETHARIA, W. R., 'Desain Kapal Prototype Armada Semut Tuna Pole and Line Material Fibreglass Reinforced Plastic (FRP) Untuk Beroperasi di Perairan Pulau Ambon'. *Laporan Akhir Program Vucer, Direktorat Pembinaan Penelitian dan Pengabdian Masyarakat, Dirjen DIKTI, DEPDIKNAS*, 2003.
11. KUSUMA, C. P. M. ; BOESONO, H. ; ARISTI, D.; PURNAMA, F., 2014, 'Analisis Hasil Tangkapan Ikan Teri Dengan Alat Tangkap Bagan Perahu Berdasarkan Perbedaan Kedalaman di Perairan Morodemak', *Journal of Fisheries Resources Utilization Management and Technology, Volume 3, Nomor 14*, 2014.
12. LAMB, T., 'A Ship Design Procedure', *Journal of Marine Technology, Vol. 6, No. 4, October 1969*, 1969
13. PURBAYANTO, 'Performa Selektivitas Alat Tangkap Bagan Rambo di Perairan Barru Selat Makasar', *Penelitian Perikanan Makasar*, 2004
14. SUDIRMAN dan MALAWA, A., 'Teknik Penangkapan Ikan', *Rineka Cipta, Jakarta*, 2004
15. SUDIRMAN, SAPRUDDIN, dan SUHARMAN, 'Hubungan Antara Kecerahan Perairan dan Kecepatan Arus dengan Hasil Tangkapan dan Pengoperasian Bagan Rambo di Selat Makassar', *Jurnal Ilmiah Sorihi, 1 (5)*, 2006
16. SUDIRMAN dan NESSA, N. N., 2011, 'Perikanan bagan dan Aspek Pengelolaannya', *Penerbit Universitas Muhammadiyah Malang*, 2011
17. SUHERMAN, A. dan ARISTI, D. P., 'Penggunaan Lampu Merkury Dalam Perikanan Mini Purse Seine di Jepara', *Laporan Kegiatan Nomor 31/2005, Fakultas Perikanan dan Ilmu Kelautan, Universitas Diponegoro, Semarang*, 2005.
18. TADJUDDAH, M., 'Pengembangan Sistim Nelayan Apung', *Fakultas Perikanan dan Ilmu Kelautan, Institut Pertanian Bogor*, 2009.
19. USEMAHU, R. dan TOMASILA, L., 'Teknik Penangkapan Ikan', *Departemen Kelautan dan Perikanan, Ambon*, 2004
20. WATSON, D. G. M., 'Practical Ship Design', *Elsevier Ocean Engineering Book Series, Volume I, ELSEVIER*, 1998

8. AUTHORS BIOGRAPHY

Wolter Roberth Hetharia, Currently holds the position of lecturer at Faculty of Engineering, Department of Naval Architecture, Pattimura University. He is responsible for the courses of ship design and resistance. Several papers have been published concerning ship resistance and optimization and fishing boats. He also presented some papers at RINA conference and other international conferences.

Fella Gaspersz is the lecturer at The Faculty of Engineering, Department of Naval Architecture, Pattimura University. She is responsible for the course of ship design and resistance. Several papers have been published concerning ship design and resistance.

Richard Latuhihin is the lecturer at The Faculty of Engineering, Department of Naval Architecture, Pattimura University. He is responsible for the course of ship construction. He present several papers concerning ship construction at local seminars.

RESISTANCE CHARACTERISTIC OF SUBMERGED PROJECTILE WITH BOW VARIATION BASED ON HULL ENVELOPE COEFFICIENT USING STEADY RANS SIMULATION

Yanuar, Ibadurrahman, and **Syifa Alfiah A. P.**, Universitas Indonesia, Indonesia

SUMMARY

Low resistance and good hydrodynamic properties are one main purpose of a submerged object design. The fore body is a major concern because it not only serve as a space for laying an acoustic equipment, but it determines the flow toward the hull. Steady RANS simulation was conducted to examine the influence of the bow shape to the surrounding flow on submerged projectile-shaped object without appendages or propulsion. The fore body is varied based on the Hull Envelope equation. The result shows the model with nf=2.75 is the optimum design according to the hydrodynamic efficiency coefficient despite the model with blunt form (nf=1) has a lowest drag because it has smallest wet surface area. Pressure coefficient distribution on the fore body span and the pressure

1. INTRODUCTION

Resistance is an important consideration in a first phase to designing a moving object since it directly affects speed, power requirement as well as dynamics of the object. Exceptionally to an object in submerged, the resistance has major influence to its top speed, accoustic signature, and hence its endurance. The total resistance of a surface vessel in calm water condition can be seperated into two elements: viscous resistance and wave resistance. Definitely, there are no wave resistance for a deeply submerged object; the viscous resistance is the only component of drag and is divided into two elements: skin-friction resistance and form (viscous pressure) resistance [1].

One way to reduce these resistance in an initial design stage is considering the hull form of the vehicle regarding to its hydrodynamic aspect [2]. Generally, the hull of a submerged object or submarine consists of three sections: fore body, middle body, and aft body [3]. The formula "Hull Envelope" as the first developed equation for a pure tear drop shape is essential in the concept design stage [4]. Another equation as an extensive hull form equation of a submerged object or submarine is used by other designer but still been associated with the tear drop shape [5]. The fore body or bow of a submerged object plays some roles on its hydrodynamic aspect, especially in order to determine a stagnation point (location and pressure). Accordingly, it forms boundary layer on the overall of the body. Moreover, the bow shape is not only affecting the resistance of the submerged object, but it remarkably influences behavior of fluid flow around it, thus controling effectiveness of the sensitive equippment in it—e.g., an accoustic equipment.

To know how a flow field arround a submerged object behaves, constructing a scale model and conducting a series of flow experiments are one method to achieve that. Nevertheless, this method is time consuming, expensive, yet well-established approach. Computational Fluid Dynamics (CFD) can be an alternative method since it gained increasingly favour in recents decades with the exponential increase of computing power and improvements in numerical algorithms; it can straightforward constructs a virtual towing tank or wind tunnel [6, 7].

The wind tunnel observation of submersible vehicle showed the flow separation for bow shape of the submarine is closer to the bow tip than the symmetric bow shape at high angle of attack [8]. A wind tunnel experiment of a high speed underwater object demonstrated hydrodynamic performance can be improved by having ellipsoidal nose profile [9]. Furthermore, an experimental of submerged spheroid had been conducted to examining the influenced of body form to the resistance components [10]. Over and above that, a CFD approach showed an elliptical submarine nose shape is a good option for a submarine design [11].

The investigation on a moving submerged object commonly focus on identifying body form to achieving an optimum design, specially the fore body. Work by using 2D steady state flow (Reynold Averaged Navierr-Stokes) RANS with a variant of the fore body shape and speed range is well underway; a commercial CFD software, ANSYS Fluent is used in this study. Objective of this paper is to examine the characteriscs of resistance and flow influenced by the bow form. As a result, it can help a designer at a concept design phase. The scope of this study focuses only on the fore body form of a projectile-like object on a deeply submerged condition where the the only considered important of the drag component is skin friction resistance and viscous pressure resistance.

2. PROBLEM DEFINITION

2.1. THE FORE BODY

The foremost aim of a bow design is to control its surrounding flow to ensure (if possible) getting a laminar flow over the forward sonar array or any sensitive equipment, also it has to extend as far as possible. The fore body is formed by revolving an ellipsee about its major axis and is described by the "Hull Envelope" equation:

$$r_{xf} = \frac{D}{2}\left[1 - \left(\frac{x_f}{L_f}\right)^{n_f}\right]^{\frac{1}{n_f}} \quad (1)$$

where r_{xf} is the radius at x_f distance in the x-direction from the rearmost part of the fore body as shown in Figure 1; L_f is the length of the fore body, D is the hull diameter, and n_f is a coefficient which defines fullness of the fore body; if nf=1, the bow shape is a conical form, and if nf=2, the bow shape is an elliptical form.

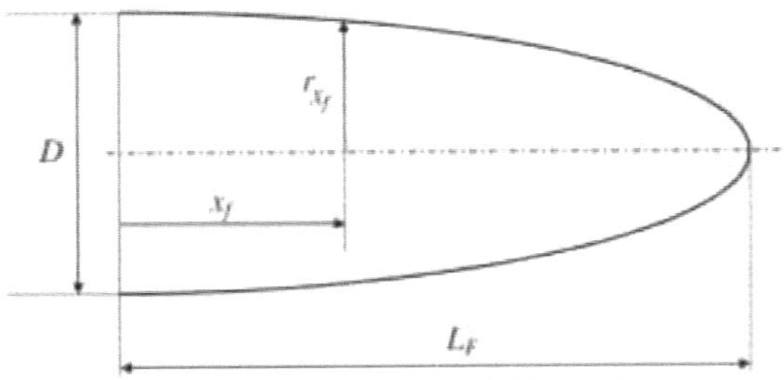

Figure 1: Schematic of the Fore Body [1]

2.2. GOVERNING EQUATIONS

Numerical simulations were performed with a commercial CFD software, Ansys Fluent. The flow around the vehicle is modeled using 2D incompressible flow, steady-state Reynolds Averaged Navier-Stokes (RANS) equations:

Continuity

$$\frac{\partial}{\partial x_i}(\bar{\rho}\bar{u}_i) = 0 \quad (2)$$

Momentum

$$\frac{\partial}{\partial x_j}(\bar{\rho}\bar{u}_i\bar{u}_j) = -\frac{\partial \bar{p}}{\partial x_i} + \frac{\partial}{\partial x_j}\left[\mu\left(\frac{\partial \bar{u}_i}{\partial x_j} + \frac{\partial \bar{u}_j}{\partial x_i} - \frac{2}{3}\delta_{ij}\frac{\partial \bar{u}_l}{\partial x_l}\right)\right] + \frac{\partial}{\partial x_j}(-\bar{\rho}\overline{u_i'u_j'}) \quad (3)$$

Where $\bar{\rho}$ is mean density, $\bar{p}$ is mean pressure, μ is the molecular viscosity and $(-\bar{\rho}\overline{u_i'u_j'})$ is the Reynolds stresses. The challenge for acquiring a high quality solution by using the RANS equation is to model the Reynold stress tensor with satisfactory; it is done by using a turbulence model. K-ε model is choosen in this examination with consideration that it has good compromise between robustness, computational cost and accuracy [12]. Turbulence kinetic energy by the model is written as:

k equation

$$\frac{\partial}{\partial t}(\rho k) + \nabla.(\rho u k) - \nabla(\rho D_k \nabla k) = G_k - \frac{2}{3}\rho(\nabla.u)k - \rho\epsilon + S_k \quad (4)$$

Energy dissipation rate (ε)

$$\frac{\partial}{\partial t}(\rho\epsilon) + \nabla.(\rho u\epsilon) - \nabla(\rho D_\epsilon \nabla\epsilon) = \frac{C_1 G_k \epsilon}{k} - \left(\frac{2}{3}C_1 + C_{3,DT}\right)\rho(\nabla.u)\epsilon - C_2\rho\frac{\epsilon^2}{k} + S_\epsilon$$

Finite volume method is employed to obtain a solution of the spatially governing equations. Coupling between pressure and velocity fields is achieved using a SIMPLE technique. A second-order upwind scheme is used for the convection and a central-differencing scheme is used for the diffusion terms.

The turbulence model has restrictions on y^+ value at the wall. The principal of the wall function is to 'bridge' the viscous flow near a wall and link the solution variables at the wall-adjacent cells to the corresponding quantities at the wall. The non-dimensional wall parameter y+ is defined as:

$$y^+ = \frac{\Delta y}{\nu}\sqrt{\frac{\tau_w}{\rho}} \quad (6)$$

where Δy is the distance from the first computational node to the wall and the subscript w denotes the wall properties.

3. NUMERICAL SETUP

3.1. MODEL DESCRIPTION

The investigated object in this paper is a submerged axis-symmetric body similar to torpedo or bullet-shape object, or even a small submarine without any appendages. The study is focusing on the bow shape effect to its resistance. This paper examines 16 bow models: 5 model of general fore body for validation and 11 models of nf-based fore body referring to the "Hull Envelope" equation. A Simplified 2D model is chosen in the examination to prove the effectiveness of CFD calculation for an uncomplicated axis-symmetric form in a steady-state condition. All part of the submerged body excluding the fore body is constant. The bow varies in each model; the aft body is conical and the middle body is cylindrical (see Figure 2). To provide an equal hydrodynamic conditions, all dimension's components are kept in constant as described on Tabel 1.

Table 1: Specification of the models

Properties	Symbol	Value	Units
Length of the model	L	6	m
Fore body length	L_f	3	m
Middle body length	L_m	1	m
Aft body length	L_a	2	m
Length/ Diameter ratio	L/D	6	-
Distance in the x-direction	x_f	0.5	m
Test velocity	v	2 – 50	knot

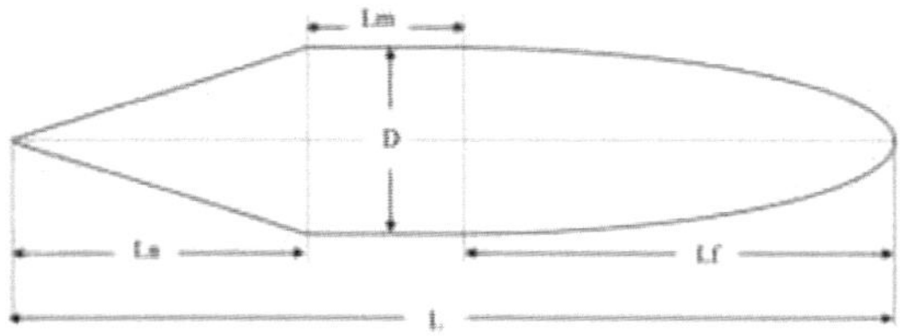

Figure 2: Hull parameter definition

Although the bow shape is different in every model, the fore body length is relatively unchanged; this helps to evaluate its hydrodynamic effect, specifically on the bow. However, every model has different surface area because its fore body variations affect the shape of each model as can be seen on Tabel 2; surface area as well as bow arc become larger for a bigger value of *nf* where the largest *nf* has a stubbed-like bow shape.

Table 2: Specific description of the models

Model	Bow shape	Bow area (m²)	Bow arc (m)
1	conic	1.500	6.083
2	ogive	1.914	6.103
3	elips	2.356	6.225
4	spherical	2.893	6.571
5	stubbed	2.996	6.914
6	$nf = 1$	1.500	6.080
7	$nf = 1.5$	2.107	6.212
8	$nf = 2$	2.361	6.224
9	$nf = 2.25$	2.490	6.295
10	$nf = 2.5$	2.511	6.335
11	$nf = 2.75$	2.618	6.343
12	$nf = 3$	2.717	6.430
13	$nf = 3.5$	2.841	6.576
14	$nf = 4$	2.964	6.717
15	$nf = 4.5$	2.972	6.762
16	$nf = 5$	2.993	6.766

There are 5 bow profile of the general bow shape for validation. The first model has a conic bow form which is uncommon in any modern submarine or submerged object because it offers insufficient amount of space for accoustic unit or other vital component of navigation. The ogive-shaped and ellips-shaped bow are applied on the novel underwater projectile with taking cavitation around its body [13]. The spherical bow shape is an ordinary form used by modern submarine or any torpedo-shaped projectile. Afterwards, the stubbed form has small resistance in surface navigation, but has large resistance in submerged mode of navigation [1].

The remaining profile of the fore body is the *nf*-based form which is designed based on the "Hull Envelope Equation" varying from nf=1 to nf=5; nf=1 is a conical form, nf=2 is an elliptic form, and so on. Increasing the *nf* value is equivalent to the model surface area along with its bow arc

3.2. BOUNDARY CONDITIONS

Boundary condition for numerical simulation of the model is illustrated in Figure 3. The inlet boundary is positioned at 2 body-length upstream with an inflow velocity from 2 to 50 knots; a pressure-outlet condition is defined as 4 body-length downstream. Free slip wall boundary condition is applied to the remaining walls with 1 body-length away from the model. Besides, no slip boundary condition is applied to the hull model.

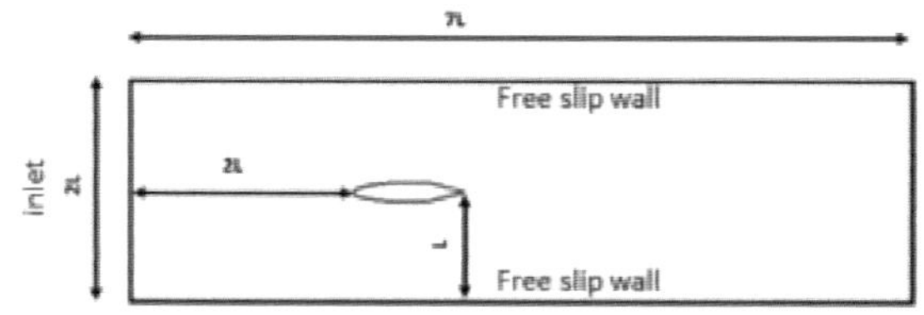

Figure 3: Boundary condition of the model

3.3. GRID GENERATION

Overall grid used in simulations is a structured hexahedral mesh consisting two parts: domain of the flow and the model wall. The flow domain employed a mesh sizing with a face meshed treatment; the model wall used a refinement treatment of the mesh with a refinement number equal to 3.

Table 3: Grid sensitivity

Mesh	Grid Quantity	y^+
Finest	5×10^5	3.57
Fine	4×10^5	3.37
Fair	3×10^5	2.83
Medium	2×10^5	3.45
Coarse	1×10^5	4.60
Rough	5×10^4	5.01

Before conducting a numerical simulation, sensitivity of the solution to the resolution of the grid is determined. Based on the base grid (rough), a series of refined grids were generated; it was refined with a ratio of 0.5 in every step of refinement as been described in Table 3.

Solution of the initial grid to the finest grid were compared to check any result variation based on the grid refinement as can be seen in Figure 4. The mesh convergence test is carried out by focusing on longitudinal force; repeated convergence was assumed through normal residuals. Form the graph, the solution changes significantly from the initial grid to the fair grade of the grid. Whereas, minor difference occurs only from the fair to the fine grid as well as the finest, and hence the finest grid was chosen for the simulation.

Beside the grid density, y^+ value as an important parameter is also related to the accuracy of the numerical prediction; the y^+ is varied from 3.57 (the finest mesh) to 5,01 for the coarsest. Therefore, the choosen grid has a y^+ value about 3.57.

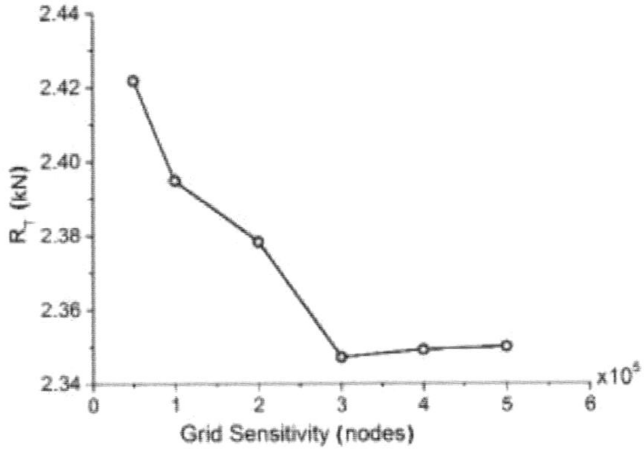

Figure 4: Grid independence

4. RESULT AND DISCUSSIONS

4.1. ITERATIVE ERROR

Stability of all equations is one parameter to achieve a convergence; at this point, the major flow features should be established. Figure 5 depicts a residual history of the simulation. Generally, a decrease in residuals by three orders of magnitude indicates at least qualitative convergence [12]. Further, solution accuracy depends on the appropriateness of the physical models and the boundary conditions.

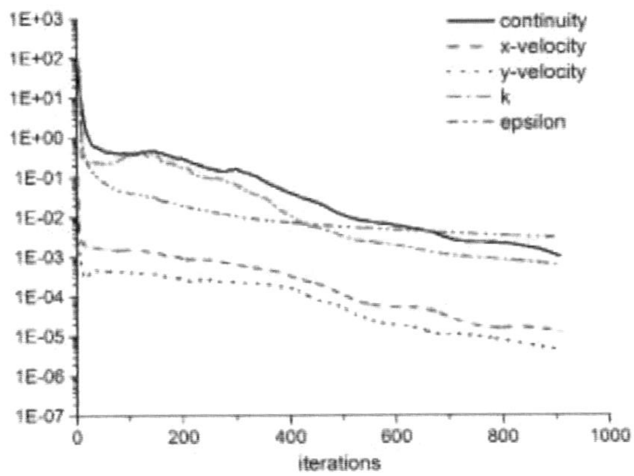

Figure 5: Residual history

4.2. VALIDATION

The validation proccess is employed by compared the numerical result with similar research as been depicted in Figure 6. The benchmark datas are produced from the result of 3D model numerical simulation with similar properties and speed test of 10 m/s [11]. From the graph, the value of total resistance coefficient (*Ct*) has no significant difference; these slight discrepancy between two simulations caused an error by an average 11,1%. The total resistance coefficients of the result are slightly higher than the reference. However, both datas have a similar tendency. Even this examination uses the 2D model as a simplified tools for designer, its numerical setup is likely good enough to continue for simulations.

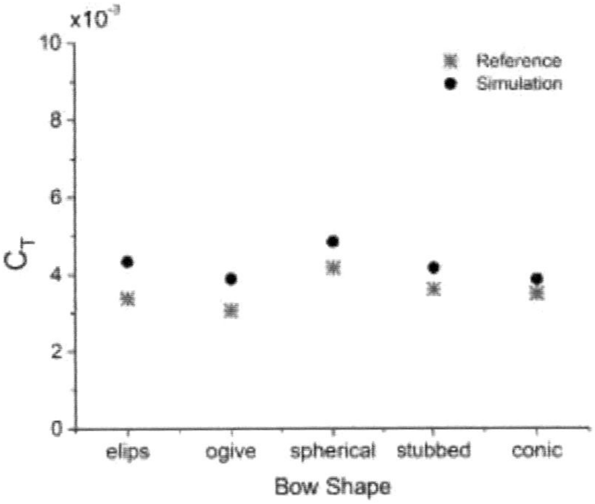

Figure 6: Validation of models with general fore body

4.3. NUMERICAL SIMULATION OF MODELS WITH *nf*-BASED BOW FORM

Designing a submerged object with fore body related to nf value is directly impacting its wet surface area and consequently is affecting its total resistance without considering its speed and another consideration. Coefficient of resistance is presented in the results instead its resistance value because it properly represent its resistance characteristics. Besides, the non-dimensional parameter is used so that each model can be properly compared all together.

Numerical simulation of model with nf=1 to nf=5 have been conducted by the finest quadrilateral nodes using 1000 iteration limit and the convergence criterion. The simulation ran for 25 speeds variation from 2 to 50 knots. From the simulation, value of total resistance coefficient (C_T), skin friction resistance coefficient (C_F), as well as viscous pressure resistance coefficient (C_P) can be acquired. These parameters is used to observe some resistance characteristics of submerged object.

Total resistance coefficient describes a general aspect of resistance characteristic of the model. Figure 7 shows the total resistance coefficient based on its Reynold Number for every model variation. As have been described before, the *Ct* on all range of *Re* becomes greater for the bigger *nf*-value model. The higher wet surface area produces higher skin friction resistance which is the biggest component of the total resistance. On that account, submerged model with *nf*=1 has the smallest total resistance coefficient where the model with highest *nf*-value crowned all models.

The pressure coefficient (C_P) depends on hull shape and consequently is influenced by the flow velocity around it. Although it accounts as the minor aspect to the total resistance of submerged object, it needs to be contemplated in order to ensure the flow phenomena around the hull, especially the bow. Figure 7 portrays pressure distribution on the models along from the forebody to the midbody.

Distinction of the trends can be seen after the tip of bow section owing to the fact that the different of the bow form based on the *nf*. Negative and positive value of C_P depends on fluid flows arround it. Generally, the blunt shape (approach to *nf*=5) has a great negative distribution of C_P, although conversely the sharp form (approach to *nf*=1) has a positive distribution.

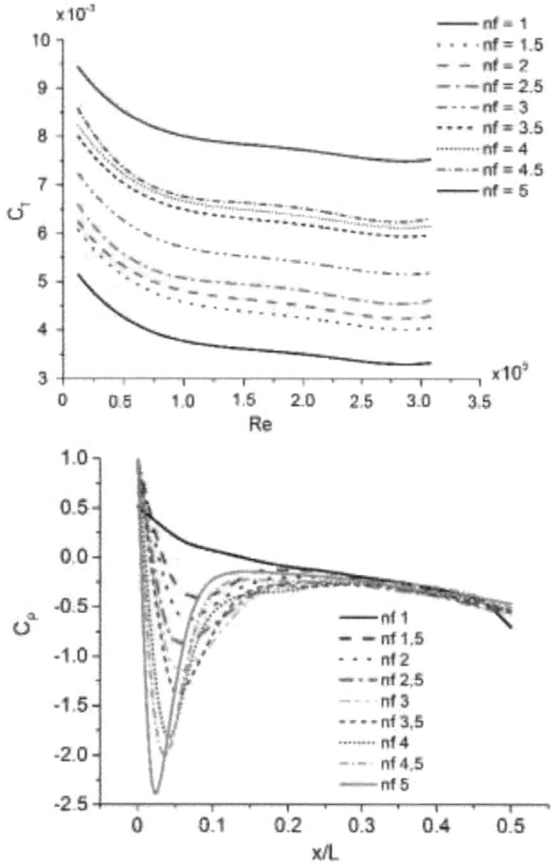

Figure 7: Total resistance coefficient (above) and pressure coefficient distribution (below) of models with *nf*-based bow form

4.4 NUMERICAL SIMULATION OF THE ELLIPSE MODELS

There are some parameter to determine the optimum bow shape for a submerged object: minimum resistance, minimum flow noise spesifically arround acoustic sensors, and fore body arrangement. According to the result, *nf*=1 or similar to the pointed shape has least space for bow arrangement even though it has the least resistance. However, the ellipse model is considered to be the realizeable form that has less resistance in addition to achieveable proper area of the fore body. Therefore, models with *nf* between 2 to 3 with ratio of 0.25 are been investigated.

As can be seen in Figure 8, coefficient of total resistance of the models has similar common tendency for the entire Reynold number span and along the *nf* variance. Again, models with a larger wet surface area (*nf*-based) has a greater resistance. The pressure coefficient shows the greater *nf* has higher pressure coefficient distribution. The tendency has similar reason to the whole models, yet it reports in a detail of alteration.

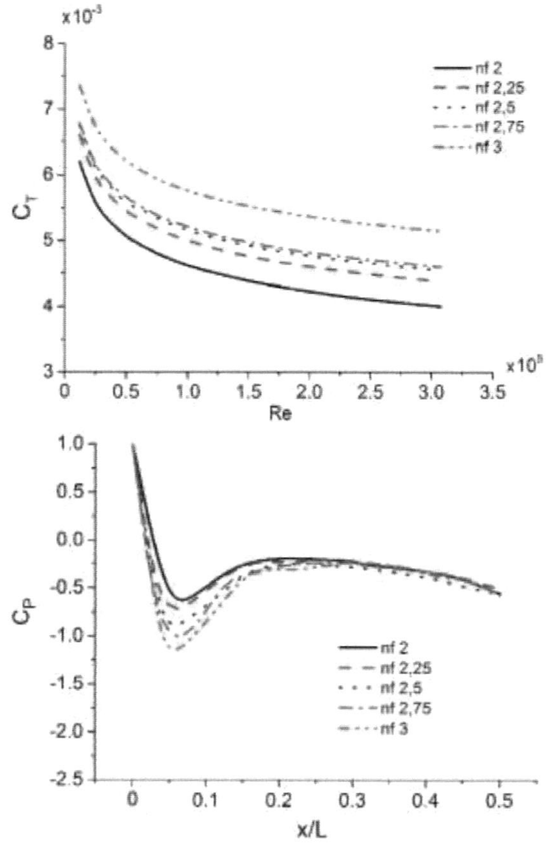

Figure 8: Total resistance coefficient (above) and pressure coefficient distribution (below) of the ellipse models

To determine an optimum bow design, a bow shape that has maximum area with minimum resistance, this study accounts a non-dimensional parameter designated as the hydrodynamic efficiency coefficient (η) and is written as:

$$\eta = \frac{A_{bow}}{arc_{bow}{}^2 x\ C_T} \quad (7)$$

where A_{bow} is the bow area, arc_{bow} is arc length of the fore body and C_T is the coefficient of total resistance. The considerable value of this non-dimensional parameter indicates the form is an optimum design.

The comparison of the hydrodynamic efficiency coefficient (η) of all models is shown in Figure 9. From the graph, models with ellipse form and similar to it (from *nf*=2 to *nf*=3) have a goodly amount of the hydrodynamic efficiency coefficient compared to the other. Thus, according to the result, the projectile model with *nf*=2.75 is the optimum form; it has a sufficient bow area while maintaining a less drag.

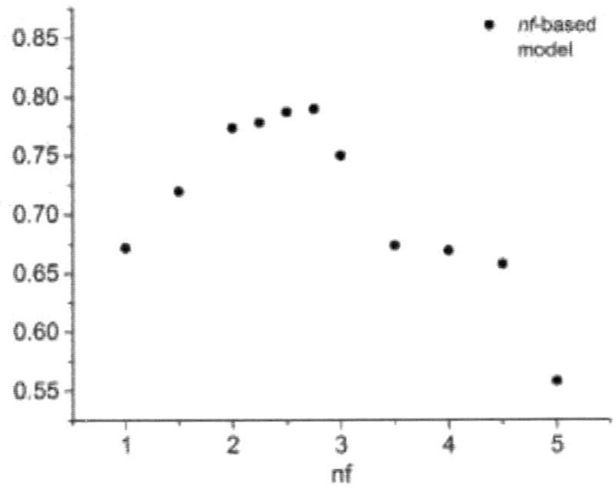

Figure 9: Comparison of hydrodynamic efficiency coefficient for all models

In Figure 10, pressure contour and fluid flow (velocity) vector around the model with *nf*=2,75 is visualized. The pressure difference along the bow is influenced by its contour. The pressure difference on the fore body is slighter for more pointed bow, and vice versa. By increasing the *nf* coefficient (more blunt), the pressure point on the tip of the fore body becomes more extensive.

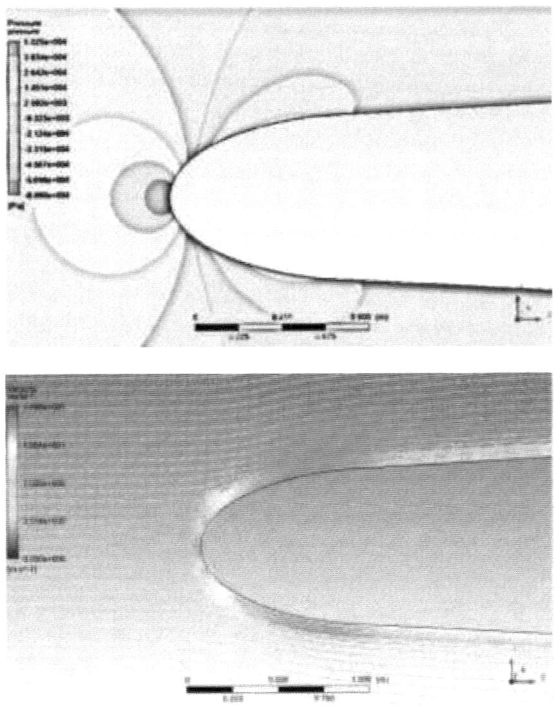

Figure 10: Pressure contour (above) and velocity vector (below) of the model 11 (*nf*=2.75)

From the velocity vector visualization, color of the flow vector represents its velocity. There are a slight flow separation after one third of the fore body length from the front end. After the shock (high pressure) at the front end, the flow is increasing rapidly, thereby making the fluid pressure decreased that leads the flow to separate from the hull. Subsequently, the flow is slowly decreasing, yet its separation is going more distant gradually.

5. CONCLUSION

Design of a submerged object depends on its hydrodynamic aspects, and importantly achieving less drag; its fore body is one of critical concerns since a flow of fluid (toward it) after reaching the bow will be affected, and consequently influenced the remaining body. In this paper, CFD simulations of 2D objects have been conducted to examine the issue with clarity. The result is able to provide a good estimate of the resistance characteristics of the submerged projectile models.

A projectile model with *nf*=1 has lowest drag since its wet surface area is smaller than the other model. However, the ellipse model (spesifically a model with *nf*=2.75) is an optimum design according to the hydrodynamic efficiency coefficient. The flow velocity is increased at one third of the bow length after the high pressure on the front end when the fluid is displaced by the bow head. Thereafter, the rising velocity of the flow makes the fluid pressure dwindled, and thus creating a separation.

The optimum design by the result of this study is suitable for the projectile-shaped object. Any excluding consideration on this paper may potentially alters the result. This study needs to be improved by using a more established 3D setup, more advanced gird resolution as well as considering another turbulence model and additional validation. Furthermore, involving additional analysis such as a relation of the fluid around the bow to the propulsive efficiency is worth to do.

6. REFERENCES

[1] M. Renilson, *Submarine Hydrodynamics*. Australia: Springer, 2015.

[2] P. N. Joubert, *Some aspects of submarine design. Part 1, Hydrodynamics (U) / P.N. Joubert*. Fishermans Bend, Vic: DSTO Platforms Sciences Laboratory, 2004.

[3] R. Burcher and L. J. Rydill, *Concepts in Submarine Design*. Cambridge: Cambridge University Press, 1994.

[4] H. Jackson, "Fundamental of Submarine Concept Design," *SNAME*, vol. 100, pp. 419-448, 1992.

[5] Y. N. Kormilitsin and O. A. Khalizev, *Theory of Submarine Design*: Riviera Maritime Media, 2001.

[6] R. Fedor, "Simulation of a Launch and Recovery of an UUV to an Submarine," *KTH Marina System*, 2009.

[7] S. Toxopeus, "Viscous Flow Calculation for Barehull DARPA SUBOFF submarine at Insidence," *International Shipbuilding Progress*, vol. 55, pp. 227-251, 2008.

[8] A. Saeidinezhad, A. A. Dehghan, and M. Dehghan Manshadi, "Experimental investigation of hydrodynamic characteristics of a submersible

vehicle model with a non-axisymmetric nose in pitch maneuver," *Ocean Engineering,* vol. 100, pp. 26-34, 2015.

[9] K. Suman, Rao, D. N., Das, H. N., Kiran, G. B. , "Hydrodynamic performance evaluation of an ellipsoidal nose for a high speed under water vehicle," *JJMIE,* vol. 4(5), pp. 641-652, 2010.

[10] C. Farell, Guven, Oktay, "On The Experimental Determination of the Resistance Components of Submerged Spheroid," *Journal of Ship Research,* vol. 17(2), pp. 72-79, June 1973.

[11] M. Moonesun, Korol, Y. M., Nikrasov, Valerie, A. Ursalov, Alexander, Brazhko, A. , "CFD Analysis of the Bow Shapes of Submarines," *Journal of Scientific and Engineering Research,* vol. 3(1), pp. 1-16, 2016.

[12] ANSYS, *ANSYS FLUENT 12.0: Theory Guide*: ANSYS, Inc, 2009.

[13] C. C.-j. Ying, LU. Jia-yi, CAO. Xin-CHEN. , "Application of Quadratic and Cubic Turbulence Models on Cavitating Flows Around Submerged Objects," *THEORETICAL & APPLIED MECHANICS,* vol. 24(6), pp. 823-833, 2012.

7. AUTHORS BIOGRAPHY

Yanuar holds the current position as Head of Ship Design Laboratory at Department of Mechanical Engineering, Universitas Indonesia. He is responsible for researches in fluid dynamics and properties. His previous experience includes includes resistance analysis of multihuls, drag reduction on ships, concept design of a conventional submarine, etc.

Ibadurrahman and **Syifa Alfiah A. P.** are a master student and a graduated student from Department of Mechanical Engineering, Universitas Indonesia.

ANALYSIS OF CONVERGENT AND DIVERGENT-CONVERGENT NOZZLE OF WATERJET PROPULSION BY ANSYS FLUENT SIMULATION

Muhammad Arif Budiyanto*, Department of Mechanical Engineering, University of Indonesia
Jansen Novri, Department of Mechanical Engineering, University of Indonesia

***Corresponding author**: arif@eng.ui.ac.id

SUMMARY

Waterjet propulsion is type of propulsion system which is used widely by ship with high speed operation and sailing in shallow waters. This system is constructed by pump and impeller that impels water passing through inlet tube. Transport energy is occured from pump to water, due to rotation of impeller. By this condition, the stream has more fluid energy for giving thrust force. Then, water stream passes thorugh the outlet tube called nozzle. Reduction of cross sectional area in the end of nozzle increases water stream velocity, outcome increasing kinetic energy by change of momentum because of an acceleration. The work of nozzle can be measured by efficiency and force output which is used to propel ship effectively. These parameters are related to nozzle geometry, as the main factor in this paper, supported by other factors such as pump pressure and stream velocity. The analysis of these parameters is presented by simulation in Ansys Fluent. The nozzle performance has been analyses by its geometry, specifically convergent and combination shape. The efficiency of convergent nozzle is bigger than combination nozzle, with difference values within 8% up to 12%. It suggests this type of nozzle is suitable for saving energy in waterjet propulsion system.

Keywords: waterjet, nozzle, convergent, divergent, efficiency, force, Ansys

1. INTRODUCTION

Waterjet propulsion uses change of momentum for working principle [1]. It is occured because water stream passes through nozzle with reduction cross sectional area. Refer to continuity equation, the velocity of stream is increasing while passing nozzle outlet. In this condition, water has ability to propel ship since having more kinetic energy, caused by both of pump transport energy and nozzle outlet. Energy from propulsion system must be equal with total resistance in ship multiplied by efficiency factor, such as frictional, residual, and air resistance [2], so that energy that generated from machinery can propel ship. Advantages from using this propulsion system are good in maneuverability, easy installation, and no complex machine alignment [3]. However, waterjet propulsion has small total efficiency, caused by cavitation factor. It arises because design of this propulsion is worked in high rotational speed with light and compact size [4]. Beside that, cavitation may occure because of limitation of pump suction head, high system temperature, and misscellaneous factors, as example, waterjet system geometrical, heat transfer and water temperature [5]. The influence of nozzle for water stream is also related to its geometry. There are three types of nozzle geometry, those are convergent, divergent, and combination [6]. The convergent nozzle is a spout that begins large and gets smaller, an abatement in cross-sectional region and vice versa for the divergent [7]. This factor influences outlet water velocity, by reduction area and inlet velocity, and other stream characteristics that can affect nozzle efficiency. This paper is intended to know the influence of nozzle geometry to water parameters, specifically nozzle efficiency and output force that generated by convergent nozzle, as the majority type of nozzle [8], and divergent-convergent nozzle. Moreover, this paper might support effort of optimization ship waterjet propulsion system. All data parameter is obtained by simulating model in Ansys Fluent.

2. METHODOLOGY

2.1. Basic and Governing Equation

Nozzle efficiency is comparison of outlet kinetic energy with fluid energy in nozzle inlet. The equation [9] is:

$$\eta = \frac{\frac{1}{2}\rho v^2}{p_i} \quad (1)$$

where ρ is water density, v is jet velocity, measured by water stream in outlet nozzle, and p is water inlet pressure. While nozzle reduction ratio, other factor that being second basic equation is comparison between outlet diameter with inlet diameter, as shown below:

$$C = \frac{d_{outlet}}{d_{inlet}} \quad (2)$$

While force equation is refered to 2nd Newton Law, and the measure is processed by Ansys Fluent program, that using governing equation, which works by computer numerically. It consists 3 equations [10] as governing equation, are mass continuity, momentum equation, and energy conservation. The equations are using for incompressible fluid, in sequences are: [11]

$$u\frac{\partial}{\partial x} + v\frac{\partial}{\partial y} + w\frac{\partial}{\partial z} = 0 \quad (3)$$

$$\rho\frac{\partial U_j}{\partial t} + \rho U_i\frac{\partial U_j}{\partial x_i} = -\frac{\partial P}{\partial x_j} - \frac{\partial \tau_{ij}}{\partial x_i} + \rho g_j \quad (4)$$

$$\rho\frac{D(e+\frac{V^2}{2})}{Dt} = \frac{\partial}{\partial t}[\rho(e+\frac{V^2}{2})] + \nabla[\rho(e+\frac{V^2}{2})] \quad (5)$$

where term of velocity is written u, v, w, and capital terms, ρ is fluid density, in this context is 1000 kg/m^3, e is term of internal energy, x, y, z, is direction volume component, t is time, P is surface pressure, and τ is shear stress.

2.2. Simulation Procedure

The type of waterjet propulsion is focused on flush type, in which nozzle inlet is almost parallel to the intake flow [12], and nozzle located in aft section of ship.

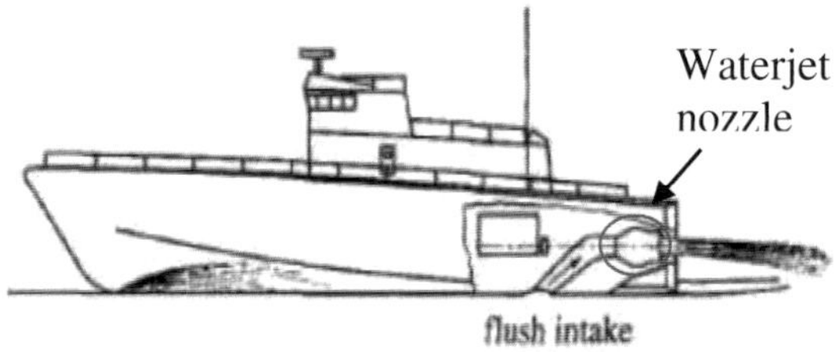

Fig 1. Nozzle position

Then, nozzle models are made as material test in Ansys Fluent. The boundary condition from models are explained by figure below.

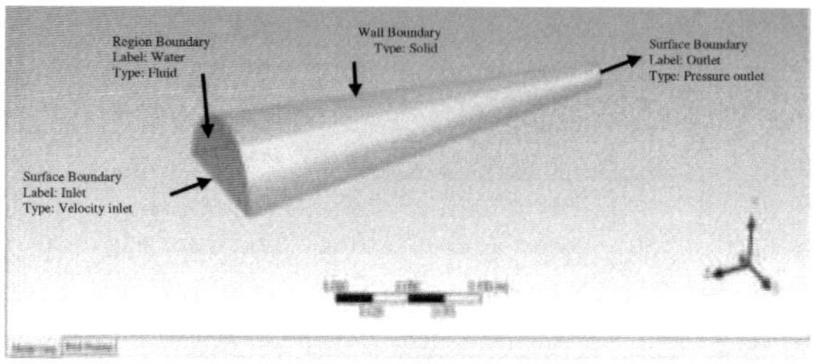

Fig 2. Symmetry model boundary condition

The model design uses dimension variation in oulet for convergent nozzle, and dimension in middle and outlet for combination nozzle. Consequently, there are three types of simulation in this paper. Varying dimension of nozzle is adressed to know difference influences of geometry, in this simulation, are diameter and position of larger diameter as divergent section in combination nozzle. The using of first two models here is aimed to know difference influence from nozzle geometry to its efficiency and force ouptut, while the third one is aimed to analyze divergence section influence at combination nozzle (divergence-convergence) characteristic, measured also from the efficiency and force output. After designing model, simulation works after setup program is fixed, conformable to journal reference as paper benchmark. All simulations are done with water velocity 6 m/s, and gauge pressure 30000 pascal. These values are obtained by trial-error methods in validation process previously. Meshing on models use medium relevance center with total about 15000 elements and 5500 nodes. Data parameter is resulted by simulation program, then those data parameters can be analyzed to achieve final results and conclusions. The working steps in this paper are shown by schematics below:

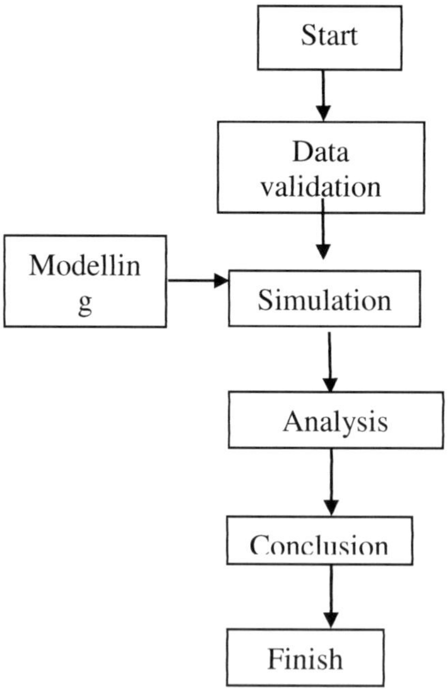

2.3 Validation Results

This paper is started by data validation referring to papers or journals that have been existed before. Validation is aimed to be setting reference for simulation, in this paper is relation between nozzle reduction area and nozzle efficiency [9].

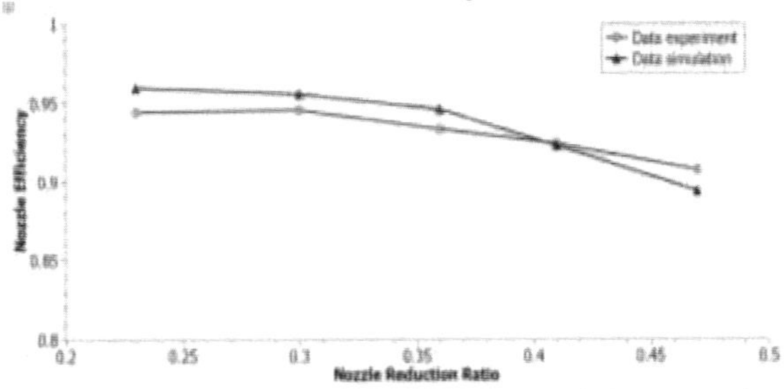

Fig 3. Comparison of relation nozzle efficiency and reduction ratio

From figure above, it is known that setting of Ansys Fluent is resemble with experimental setting, with root mean square of simulation is 0.0256, which is relative small error, refers to actual result in experimental data. After validation, the setting can be used in true simulation, that aimed firstly in introduction.

2.4. Simulation Parameter

In this paper, nozzle geometry types that used as model are convergent and combination. The dimensions of model shown by these tables below:

Table-1. Dimension of convergent nozzle

Inlet diameter (mm)	Outlet diameter (mm)
77	17.53
	23.47
	27.55
	31.82
	36.37

Table-2. Dimension of combination nozzle A

Inlet diameter (mm)	Middle diameter (mm)	Outlet diameter (mm)
77	100	17.53
		23.47
		27.55
		31.82
		36.37

Table-3. Dimension of combination nozzle B

Inlet diameter (mm)	Middle diameter (mm)	Outlet diameter (mm)
77	84.7	31.82
	92.4	
	100.1	
	107.8	
	115.5	

3. RESULT AND DISCUSSION

3.1. Nozzle Efficiency

From simulation, stream velocity and inlet pressure are obtained with error less than 0.001 or 0.1%, which informs simulation results are convergent in certain values. As shown in Fig. 4, the nozzle efficiency, refers to equation 1, is decreasing in line with increasing diameter of nozzle outlet.

Then, efficiency values from convergent nozzle are bigger with differences within 8% up to 12% than combination nozzle, and curves relatively have same tendency, because both of them shows decreasing nozzle efficiency in line increasing outlet diameter.

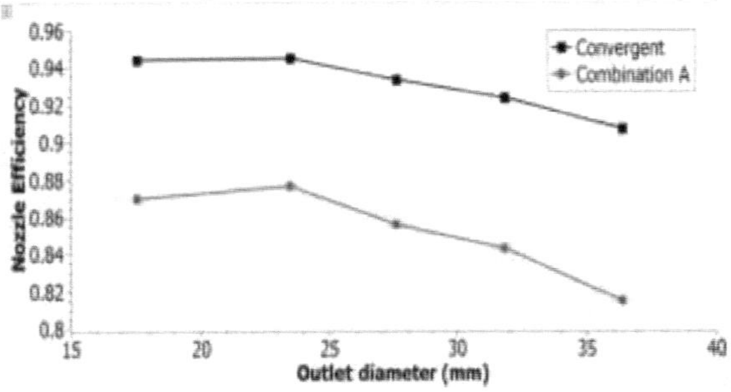

Fig 4. Relationship between outlet diameter and nozzle efficiency

In model simulation by varying middle dimension, that seems graphic below does not figure tendency of curve. For dimension in this paper, maximum efficiency is occurred in middle diameter 92.4 mm by 87%, while the lowest is 107.8 by 81%. By the result in Fig. 5, diameter in middle section of nozzle maybe not absolutely used for determining tendency nozzle efficiency, so it might have specific result in difference condition and setup simulation.

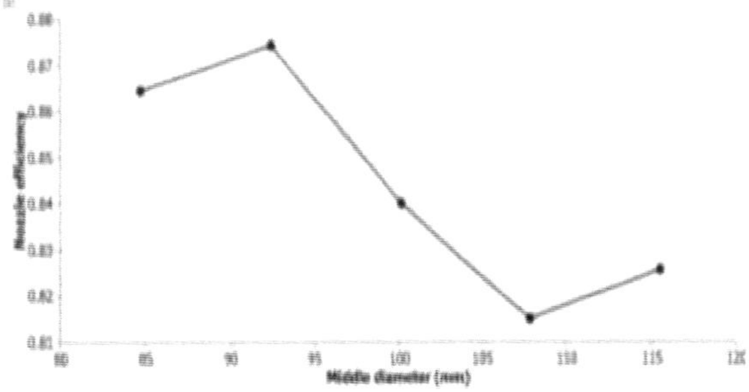

Fig 5. Nozzle efficiency in combination nozzle B

Table-4. Summary of nozzle efficiency

Number model	Convergent	Combination A	Combination B
1	0.96	0.88	0.86
2	0.96	0.86	0.87
3	0.95	0.84	0.84
4	0.92	0.81	0.81
5	0.89	0.77	0.83

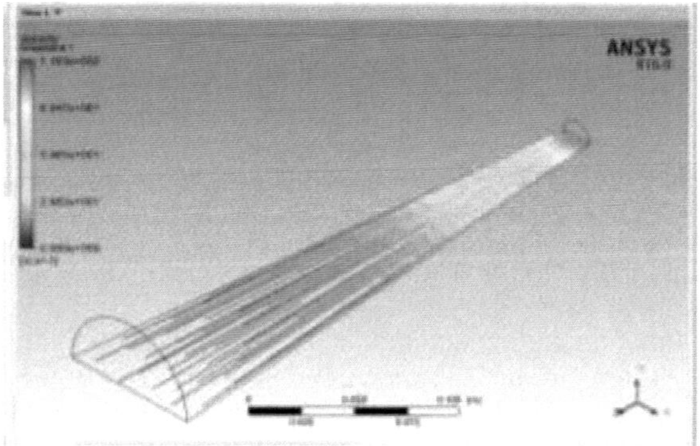

Fig 6. Velocity streamline convergent nozzle outlet diameter 17.53 mm

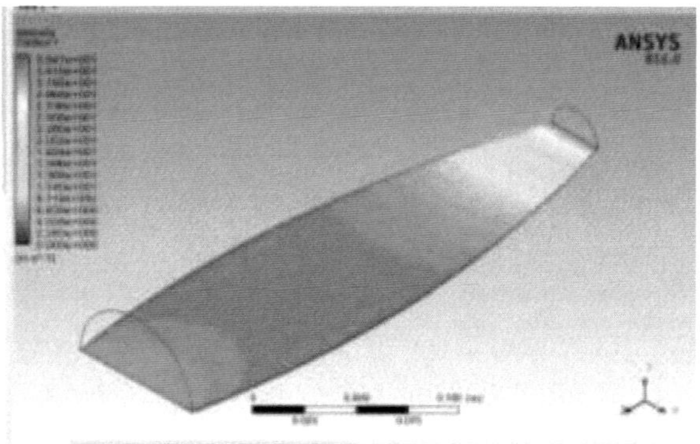

Fig 7. Velocity contour combination nozzle B middle diameter 92.4 mm

3.2. Force Output

Same with nozzle efficiency simulation, results of force output are obtained with error less than 0.001 or 0.1%, that represent values of force is technically convergent

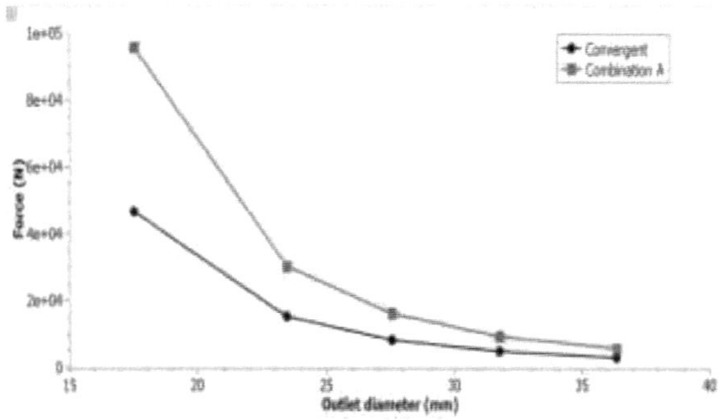

Fig 8. Relationship between force output and outlet diameter

Figure 8 indicates force output combination nozzle is 1.5 up to 2 times from convergent nozzle and Fig. 9 shows that increasing middle diameter of nozzle also followed by increasing force output.

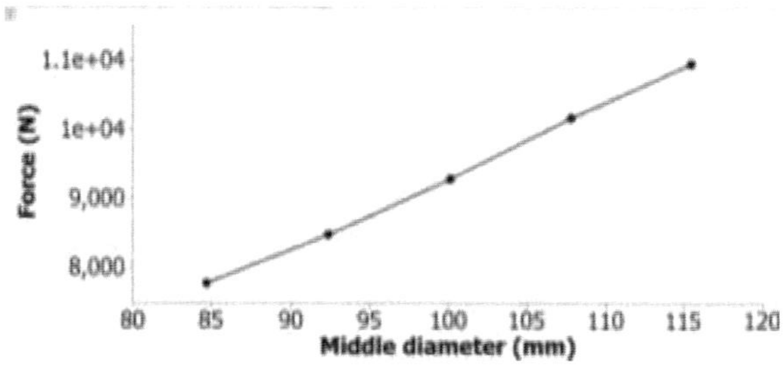

Fig 9. Force output in combination nozzle B

Figure 9 shows force output in relation to the combination nozzle middle section which obtained by the simulation. Result from simulation is plotted in graphic to figure the tendency of curve. The graphic above indicates force output is increasing parallel with increasing of middle diameter, that maximum force value is 10938 N, in 115.5 mm middle diameter. Because force output is measured by acceleration of water stream, refers to continuity equation, there is biggest difference stream velocity because of more differences dimension in the middle with outlet section, hence, bigger diameter in the middle, the acceleration is also increasing, eventually system will develop bigger force output.

Table-5. Summary Force Output

Number model	Force output (N)		
	Convergent	Combination A	Combination B
1	46546.14	95528.00	7761.02
2	15218.32	30231.00	8467.39
3	8290.73	16138.00	9276.10
4	4819.82	9258.13	10157.00
5	2932.05	5642.98	10938.00

4. CONCLUSIONS

Simulation in this paper is aimed to analyze nozzle performance that influenced by its geometry, specifically convergent and combination shape. The performance is indicated by values of nozzle efficiency, which mathematically depends on inlet pressure, density, and jet velocity, measured in outlet section, and the output force, which resulted by difference of momentum and acceleration that occurred in water stream. The summary of this simulation as follows:

1. The efficiency of convergent nozzle is bigger than combination nozzle, with difference values within 8% up to 12%. It suggests this type of nozzle is suitable for saving energy in waterjet propulsion system. Then, variation on middle diameter in combination nozzle does not given a tendency of nozzle efficiency, and it might have unique result, corresponding to nozzle dimension and condition setup.
2. The force output of combination nozzle is almost two times convergent nozzle, because of bigger diameter in middle section cause change of momentum in the outlet increasing, followed by increasing of force output value. It suggests combination nozzle may be considered to the using of waterjet propulsion system which need high acceleration, that can be obtained by extra force output, resulted by nozzle. Moreover, since bigger middle diameter of nozzle, it increases force output, which is resulted by the outlet of nozzle.

ACKNOWLEDGEMENTS

Authors would like to express our gratitude to Department of Mechanical Engineering Universitas Indonesia in the providing of research facility and publication funding

REFERENCES

[1] Wardanu, Y.S., Santoso, Agoes, Adji, S.W. 2013. Rancangan nozzle waterjet untuk meningkatkan kecepatan renang pada Tank BMP-3F (Infantry Fighting Vehicle). Teknik Sistem Perkapalan Institut Sepuluh Nopember Surabaya, pp: 139-143

[2] MAN Diesel and Turbo. Basic principles of ship propulsion. 2011. Copenhagen, pp: 10-12

[3] http://www.marinewiki.org/index.php/Waterje t_prop ulsion; accessed June 18th, 2017, 10.22 AM

[4] Arcand, Lionel, Visitor. 1966. Waterjet propulsion for small craft. Proceedings of Southeast Section Meeting Miami, Florida, pp: 8

[5] Toloian, Akbar. 2015. The cause of cavitation and prevention methods of this desctructive phenomenon in pumps. Technical Council of East Azerbaizan Province, pp: 141-149

[6] https://www.kullabs.com/classes/subjects/units/le sso s/notes/note-detail/4789; accessed June 6th, 2017, 08.57 PM

[7] http://www.chegg.com/homework-help/definitions/t he-convergent-nozzle-5; accessed June 20th, 2017, 07.02 PM

[8] Schneekluth, H., Bertram, V. 1998. Ship design for efficiency and economy. Oxford: Butterworth-Heinemann, pp: 121-128

[9] Fujizawa, Nobuyuki. 1995. Measurement of basic performances for waterjet propulsion system in water tunnel. Proceedings of the Interantional Journal of Rotating Machinery, pp: 43-50

[10] ANSYS fluent theory guide. 2013. Canonsburg

[11] Versteeg, H.K., Malakasekera, W. 1995. An introduction to computational fluid dynamics. Essex: Longman Scientific and Technical

[12] Eslamdoost, Arash. 2014. The hydrodynamics of waterjet/hull interaction. Thesis for the degree of Doctor of Philosophy, Chalmers University of Technology Gottenburg, pp: 6-7

AUTHORS BIOGRAPHY

Muhammad Arif Budiyanto, is lecturer Naval Architecture and Marine Engineering, Universitas Indonesia. His subject is energy saving technology in maritime industry by numerical analysis approach. He obtained doctor degree in Urban and Environmental Engineering Department, Kyushu University. He received B.A. in Naval Architecture and Marine Engineering, University of Indonesia, and M.A. in Mechanical Engineering, University of Indonesia.

Jansen Novri is currently as undergraduate student in Naval Architecture and Marine Engineering, Universitas Indonesia.

Seawater Ice Slurry Generator with Simetric Scraper and Propane for Fish Cooling

Intan Permata Sari, Niken Dhanitya, Fajri Ashfi Rayhan, Agus S. Pamitran*, Department of Mechanical Engineering Universitas Indonesia, Indonesia, *pamitran@eng.ui.ac.id

SUMMARY

Sea water ice slurry can be produced with cooling process in evaporator by utilizing salt content in seawater as freezing point depressing additive. Ice slurry technology can play an important role in the process of storing fish in the open sea because the quality of the fish cooled with ice slurry is much better than ice cubes. Evaporator with polytetrafluoroethylene (PTFE) coating blade and a certain clearance distance can reduce electric power and time consumption in ice slurry generation. The present study used variation of pump RPM and scraper RPM. The best production rate of ice slurry generator is 0.284 Ton per 12 hours using 10 ppt salinity with scraper RPM of 335 and pump RPM of 1064. The fastest ice slurry generation is at constant 1005 RPM of pump with scraper RPM of 357 viz. 22 minutes since sea water temperature reach 15°C.

Keywords: Seawater, fish, ice slurry, evaporator

1. INTRODUCTION

One of the essential thing in designing fast vessel is to Opportunity of development of fishery business of Indonesia have very high prospect. The economic potential of marine and fishery resources that can be utilized to boost economic growth is estimated at USD 82 billion per year. The potential of the Indonesian sea is huge. National fish stocks in 2015 increased from 6.04 tons to 7.3 tons compared to 2014 [1]. However, the utilization of fishery products in Indonesia has not been maximally overall. This is caused by various factors, among others: the ships used to catch the fish is still traditional, the cost of fishing boat repairs, the low fishermen's welfare, and the lack of utilization of latest technology in fishing and fish management. One technology developed in preserving fish is ice slurry as a cooling media for fish [2]. Ice slurry can cool the fish for 6 days [3]. It states that ice slurry can cool fish better than crushed ice.

Currently, there have been some researches on developing generators to produce ice slurry. The general difference that distinguishes the developed ice slurry generator is the type of scraper, including rod orbital type, screw, blade, etc. In the present study the generator use the type of scraper with two-symmetric blade. The experiment was performed with a two-symmetrical blade scraper installed in evaporator. The scraper is designed with PTFE coating on the blade hence it is expected to reduce friction on the scraper blade. The evaporator also has an adjustable blade size so it has a certain clearance value. The aim of the present study is to investigate the characteristics and performance of the developed ice slurry generator with some experimental conditions. The developed ice slurry generator is expected to produce ice slurry with good cooling quality and with shorter manufacturing time. The study focuses on the performance of ice slurry generators, the effect of seawater flowrate, and the effect of seawater salinity on the ice slurry generation.

2. METHODOLOGY

Stages of research conducted are to find the effect of some variables on the performance of developed ice slurry generator. The test is running with ten liters of seawater. In the present study, the production of ice slurry requires two cycles running simultaneously viz. refrigeration cycle and the seawater circulation cycle as shown in Figure 1. Temperature of ice slurry and refrigerant was measured by thermocouple and recorded by data acquisition device. Pressure of refrigerant is measured at some locations as shown in Figure 1.

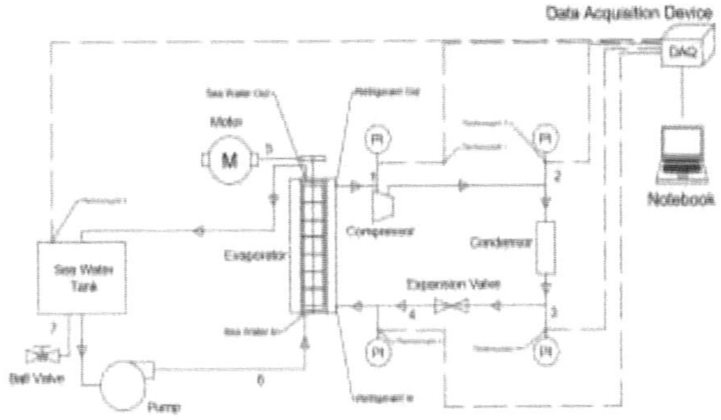

Figure 1. Schematic of Ice Slurry Generator System

An evaporator as a main part of ice slurry generator was developed in this study as shown in Figure 2. The evaporator used a working principle adapted from a concentric tube, or tube in tube heat exchanger. The refrigerant is fed from the bottom side to the larger diameter tube absorbing some heat from seawater through conductive heat transfer hence the seawater change become ice slurry. The evaporator has a clearance of 5 mm between the PTFE coated blade and the inner surface of the smaller tube in order to reduce the friction.

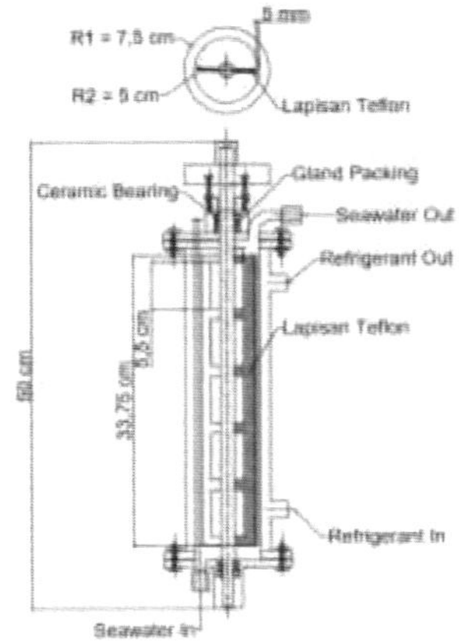

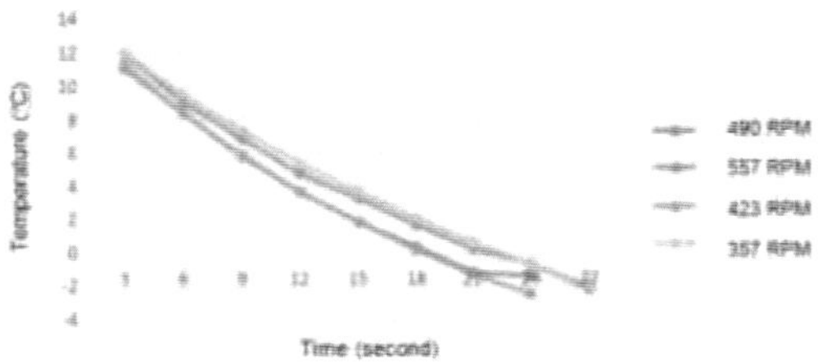

Figure 2. Schematic of evaporator

3. EXPERIMENTAL RESULTS

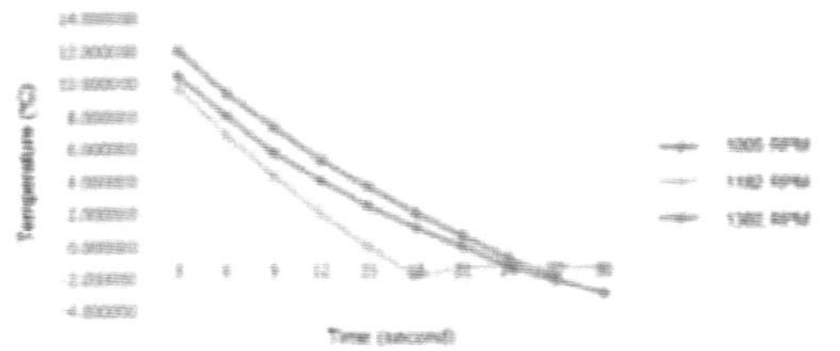

Effect of Scraper RPM on Ice Slurry Generation

A centrifugal pump is used to circulate the seawater through the evaporator. The pump works with constant RPM in order to observe the effect of scraper RPM on the ice slurry generation. Figure 3 shows that the higher scraper RPM result in longer time of ice slurry generation. High scraper RPM make the heat transfer is more evenly distributed in the evaporator and it scrapes the ice growing on the evaporator surface. Higher scraper RPM needs higher electric power consumption. The result shows that the largest electric power is 0.35 kWh at scraper RPM of 423 and pump RPM of 1242. The fastest ice slurry generation is at condition of scraper RPM of 1005 and pump RPM of 357, viz. it takes 22 minutes start from the temperature of sea water of 15°C, and then the scraper can continue work for 23 minutes until the entire sea water becomes ice slurry.

Effect of Pump RPM on Ice Slurry Generation

The function of a scraper on an ice slurry generator system is to stir up the seawater in the evaporator and to scrap the ice growth on the inner wall surface of the evaporator. In order to find the effect of pump RPM, the pump works in some test conditions with constant scraper RPM. Figure 4 shows the higher the pump RPM, the faster the sea water temperature decreases. This is due to the faster RPM of the pump the faster the sea water circulation in the evaporator, thus causing faster heat absorption.

Figure 3. Effect of scraper RPM on decreasing of seawater temperature with constant pump RPM of 1005
Figure 4. Effect of pump RPM on decreasing of seawater temperature with constant scraper RPM of 357

7. CONCLUSIONS

A study of ice slurry with developed evaporator was conducted in order to observe the ice slurry generator system performance with some experimental conditions of pump RPM and scraper RPM. The developed ice slurry generator has an optimum operation with pump and scraper RPMs of 1005 and 357, respectively. The experimental condition with higher scraper RPM result in slower ice slurry generation. The higher the pump RPM the faster the sea water temperature decreases. Ice slurry production rate increases with lower scraper and pump RPMs

8. ACKNOWLEDGEMENTS

The authors would like to express sincere gratitude to Univeristas Indonesia to support our research.

9. REFERENCES

[1] Kementerian Kelautan dan Perikanan RI, 2015. Laporan Kinerja Kementrian Kelautan dan Perikanan.
[2] A.S Pamitran, H.D. Ardiansyah, M. Novviali, 2013, A review paper of sea-water ice slurry generator and its application on Indonesian traditional fishing, Applied Mechanics and Materials, 388, p.128-132
[3] Kitamura, K. et.al., 2015, Introduction of slurry production apparatus capable of making ice from 1 wt% salinity, J Fishing Boat and System Engineering Association of Japan, 122, p.2

Effect of Geometry Characteristic of Oil Tanker's T Stiffener Profile towards Ultimate Strength under Axial Compressive Load

Firman A Nugroho1, Chandra Ardika 2
1Departement Mechanical Engineering,University of Indonesia, Depok 16424, Indonesia
2Undergraduate Student of Departement Mechanical Engineering,University of Indonesia, Depok 16424, Indonesia

SUMMARY

Future LNG field development project in Indonesia will be dominated by local buyer of the country, so that additional LNG tankers are needed for the domestic market in the future. Theoretically Dual Fuel Steam Mechanical (DFSM) or steam propulsion system has a low thermal efficiency, but more reliable than the DFDE propulsion system. On the other hand Dual Fuel Diesel Electric (DFDE) propulsion system has shortcomings in the technology that is the presence of methane slip potential for global warming with a Global Warming Potential (GWP) of 20-25 times greater than CO_2 over a period of 100 years or 20 times at intervals of 20 years, thus actually reducing the achievement of the energy efficiency of the propulsion system. In this paper, two LNG ship propulsion system that is DFSM and DFDE which is currently the most widely used by operators in Indonesia were compared technically and economically. Technically, the IMO guidelines on Energy Efficiency Design Index (EEDI) and the Energy Efficiency Operational Index (EEOI) were used as a means of comparison, while from the economical side life cycle cost (LCC) CAPEX and OPEX cost-based of the ships were analysed. Based on the technical evaluation using the guidelines EEDI and EEOI of LNG tankers by considering the slip of methane in the DFDE propulsion system, showed that the value of the achievement of EEDI and EEOI the DFSM propulsion system lower (more efficient) than the value of the achievement of the DFDE propulsion system and economics evaluation based on the LNG tankers cost life cycle (LCC) for 20 years by considering the risks of loss and penalty carriage of LNG that may occur due to the failure of the propulsion system and the BOG system, shows the DFSM propulsion system more efficient than DFDE system.

1. INTRODUCTION

1. Introduction
Ship comprises of complex structures for during the operation on the ocean, the dynamic and static loading working on the hull of the ship continuously [1]. Such condition is supported by the sub-structures for instance deck, double bottom, girder, frame, and stiffened plate. Stiffener has been understood as main sub-structure which supports the plate in common. There are several types of stiffener profile for instance flat bar, angle bar ad T bar [2]. Each of the profiles has specific characteristics for the load working on the ship has different characteristics i.e. the side hull structure has differ from the bottom structure. For instance, on the bulk carrier and the oil tanker, the angle stiffener profile is common installed on the side structure and the T stiffener profile is on the bottom structure [3]. Therefore, the geometrical profile of the stiffener has detail purpose in supporting the ship structure for varied load working on the hull of the ship.

Accordingly, the strength of the stiffener profiles becomes major concerns, especially the ultimate strength. Definitively, ultimate strength clearly describes the load-displacement curve on which characterize the tenacity of the stiffener profile under specific type of loading [4]. Such loading can be generated from different wave height in terms of three degree freedom x-abscissa, y- ordinate and z- applicate i.e. crest and trough of the wave. The sagging and hogging during the operation of the ship generally produce vertical bending load, axial compression, lateral pressure, and torsional acting on the sub-structure i.e. stiffened plate with varied geometrical stiffener profiles. The axial compressive load is chosen as a major variable on this work, for simply such stiffener profiles can be both plastically and elastically deformed, on which elaboration regarding the correlation among ultimate strength, section modulus and mass toward geometry characteristics of conventional T stiffener profile is necessary.

Several related research have been conducted, e.g. different geometry of stiffened panel i.e. Comparisons of calculations with experiments on the ultimate strength of wide stiffened panels[5], investigation on the dented wide and narrow stiffened panel towards residual ultimate strength under compressive load[6-7]. Other research on varied dimensional condition of box girder with regard to ultimate strength under torsional loading also has been carried out [8-9]. However, lack of the research regarding the mapping of effect geometrical characteristics on stiffener profile toward structural performance ultimate under specific loading left wide area of improvement. Especially on the real maximum load carrying capacity of flat bar, angle bar and T bar stiffener profile which can withstand using non-linier finite element analysis. Furthermore, this could be strong foundation for the development of Y stiffener profile as novel stiffener on the panel with regards to the effectiveness of structural performance compared to the conventional ones.

Regarding the implementation of the non-linear finite element method on the sub-structure, Paik et al [10-11] presented the nonlinear finite element method models for ultimate strength analysis of sTl stiffened-plate structures under combined biaxial compression and lateral pressure actions—Part I : plate element, and Part II : Stiffened panels. In addition, Paik [12] also elaborated the residual ultimate strength of sTl plates with longitudinal cracks under axial compression in accordance with the non-linear finite element method investigation. Do HC et al [13] carried out study on the ultimate limit state assessment of stiffened panel structures for very large ore carrier. Therefore, on this study, ANSYS Non-linear finite element employed on number of stiffener profiles which are typically found within oil tanker and bulk carrier structures to elaborate the ultimate strength characteristics, section modulus and mass with variation of geometrical characteristics of conventional T stiffener profile under axial compressive load.

2. Referenced oil tanker structural characteristics

On the present study the object the T stiffener profile on the mid-ship section of oil tanker designed by the IACS-CSR method on which refer to Badran et al[14]. Accordingly, table 1 explains the geometrical characteristics of the double hull oil tanker. Fig. 1 describes the mid-ship section of the hull ship on which the stiffener profiles on the inner and outer bottom and deck structure are the major concerns, especially on the dimension of each T stiffeners profile which different each deck, side, and bottom structure. Therefore, the thickness difference is also considered with regards to the optimum condition of moment inertia of each configuration of the stiffener profiles. This is related to the geometrical properties for a given cross-section of stiffeners i.e. Section modulus and the mathematical structural strength variable, bending stress and moment. Accordingly, on the present works, the value of section modulus is calculated using analytical approach.

The conventional stiffeners have different profile and geometry characteristics on main deck, side and bottom structure. Refer to the fig. 1, the details geometrical characteristics of side and bottom structure also are divided into inner and outer structure, respectively. Fig. 2 describes the conventional stiffener profile e.g. flat, bulb and T, on which the geometrical design referred to the international association classification societies. The T stiffeners profile known as major stiffener on the bottom, side and top deck structure, table 2 verifies the geometry dimension of 29 conventional T stiffeners profiles on the mid-sip section of double hull oil tanker, on which the variable of geometrical dimension covering the plate breadth, plate thickness, web length, web thickness, flange breadth and flange thickness. The sequence of stiffener position is started from the top outer side structure to the bilge structure at corner of bottom. It started again from the top of the inner side structure to the bottom and continued to the inner of bottom of the structure transversely and the outer bottom of the structure, and the last stiffener is on the longitudinally center-line of mid-ship section.

Table 1
Geometrical characteristics of an oil tanker

Length between Perpendicular (LBP)	238m
Moulded breadth (B)	43m
Moulded depth (D)	21m
Scantling draught, (T)	14.3m
Deadweight, (DWT) tonnes	97000

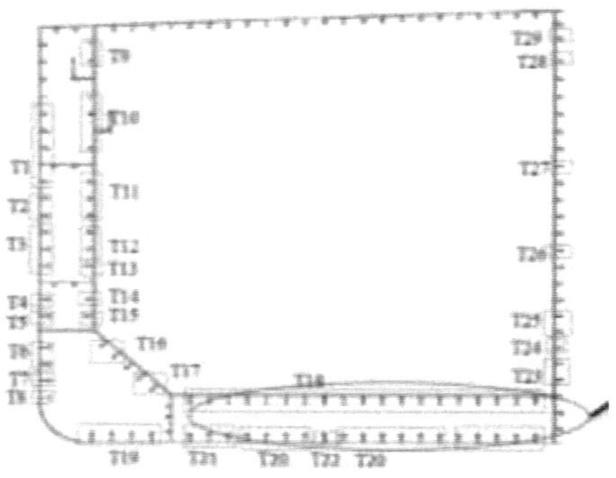

Fig. 1 Mid-ship section of double hull oil tanker.

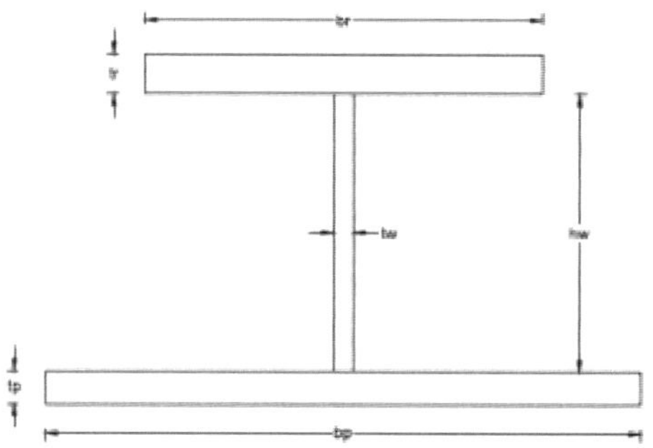

Fig. 2. Conventional T cross sectional Stiffener profiles

Table 2 Geometry of T stiffener profile of mid-ship section of oil tanker (Badran et al)

T	Plate Breadth	Plate Thickness	Web Length	Web Thickness	Flange Breadth	Flange Thickness
T1	850	15.5	300	12	150	15
T2	850	15.5	330	12	150	15
T3	850	15.5	350	12	150	18
T4	850	16.5	380	12	150	18
T5	850	16.5	440	12	150	18
T6	850	17	420	12	200	18
T7	850	17.5	420	12	200	18
T8	850	17.5	450	12	200	18
T9	850	17	600	12	175	15
T10	850	15	300	12	150	15
T11	850	15	300	12	150	18
T12	850	15	360	12	150	18
T13	850	15	370	12	150	18
T14	850	15.5	370	12	150	18
T15	850	15.5	420	12	150	18
T16	850	17	400	12	150	18
T1	850	17.5	420	12	150	18

7						
T18	850	15.5	430	12	150	18
T19	850	15.5	470	12	150	18
T20	850	18.5	450	12	150	18
T21	850	17.5	450	12	150	18
T22	850	18.5	430	15	150	18
T23	850	15.5	460	12	200	18
T24	850	15.5	450	12	175	18
T25	850	16	450	12	175	18
T26	850	15	430	12	150	18
T27	850	15	330	12	150	18
T28	850	17.5	600	12	175	15
T29	850	17.5	600	14	200	15

3. Nonlinear finite element analyses on T stiffener profile

3.1 Finite element model

According to the previous several research that showed well précised on the results of non-linear FEA, it is believed that ANSYS non-linear FEA may be the most suitable refined method to give accurate solution, thus the modeling technique applied must be well established on representing actual structural behavior associated with geometrical non-linearity, material non-linearity, boundary condition, loading condition, mesh size and so on [8, 12]. Elastic-plastic modeling of material stress-strain properties considered on this present study. The elastic-plastic large deflection behavior before and after the ultimate strength is reached, on which the elastic-perfectly plastic material model is adopted by neglecting the effect of strain hardening [8]. The accuracy of non-linear finite element study is controlled by mesh refinement. Therefore, coarse and fine mesh considered for conventional stiffener models with difference on geometrical variables. The fine mesh is 30 % denser than the coarse mesh, thus the mesh density is relative. The mesh size is selected from verification by Saddek [14] to achieve the critical moment for a simply supported beam under uniform moment.

ANSYS 15.0 student version is used to assess the ultimate strength of the conventional T stiffener profile. Fig. 3 shows the design model of conventional stiffener profile on which the geometrical and material non-linearity is considered with elastic-plastic large deflection. A bilinear isotropic elastic-plastic material model with no consideration of stain rate effects is used in the FE analysis. The material properties are Young's modulus, E = 206GPa; Poisson ratio, m = 0.3, yield stress, $Y=235N/mm^2$. The length of the stiffener is 4500mm. Solid 45 element (tetrahedral) is used to model the stiffener in finite element analysis. This element has eight nodes on each have 3 degrees of freedom: translation in the nodal x-, y- and z-direction. The solid 45 has plasticity, large deflection and strain capabilities. Accordingly, a representative of accuracy can be obtained for the actual spread of plasticity and yielding behavior of stiffener model. Considering that the total stress range is equal to double of the yield stress, the nonlinear stress-strain model is bilinear kinematic hardening (rate independent plasticity) behavior model i.e. using a simple bilinear stress-strain curve. It is also suitable for a small strain used for material that in line with the von Mises yield criterion i.e. steel. The solid 45 of FE model is fine enough to give accurate and precise result, considering the computation time savings required a proportional accuracy and effort. In addition the level of inflation should be considered for affecting the accuracy. Thus, the element size to the thickness ratio is considered.

3.2 Boundary condition and loading

Regarding the prescription of boundary condition in models, it is important to model the stiffener edge condition in appropriately, since the boundary of conventional T stiffener profile is supported by longitudinal girder and transverse frames. The boundary condition covers the T[x, y, z] which indicates the translational constraints, and the R[x, y, z] indicates rotational constraints in x-, y-, z-coordinates, respectively; a "0" indicates constraint, and a "1" indicates no constraint [12]. The representative axial compressive load working on the ship during the operation, is conducted by implementing the uniform load on the cross sectional area, meanwhile the other cross sectional area acted as fixed support, as shown in fig. 4. The given uniform load 6000 kN to investigate the longitudinal displacement of the stiffeners. Therefore geometrical difference among stiffeners, affecting the section modulus and the ultimate strength can be more elaborated.

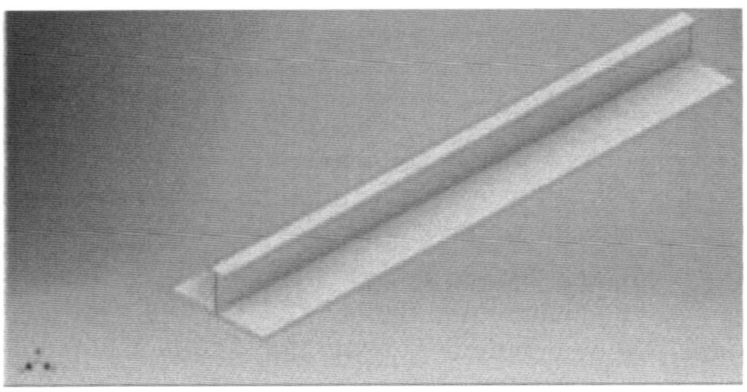

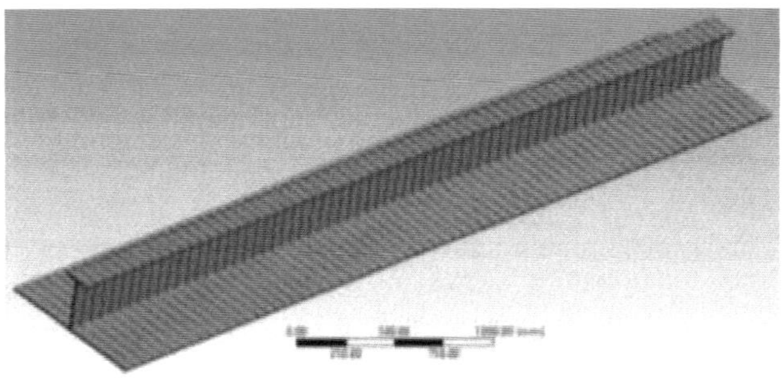

Fig.2. Design of conventional T stiffener profile
Fig. 3. Meshed of designed stiffener

Fig. 4 Longitudinal compressive load

For the purpose of this study three configurations of two LNG tanker propulsion system is used as an object of comparison. As a sample of an LNG tanker with DFDE propulsion system is divided into two types with major differences in the type of system configuration of the main engine. The detailed characteristics of LNG tankers that will be used as reference in this study are in Table 4.1 as follows:

4. Section Modulus

As the geometrical characteristics become major concerns, the affected physical variable is section modulus. Definitively, section modulus is capability of a beam to resist the form alteration of the given load. The value of section modulus is based on the geometrical design of 29 stiffeners on the table 2. Accordingly, the geometrical dimension of the stiffeners is described in fig. 5, on which varied for 29 stiffeners in terms of dimensional variable i.e. plate breadth (Pb), plate thickness (Pt), web length (Wl), web thickness (Wt), flange breadth (Fb) and flange thickness (Ft). The analytical approach was used on this work, the results can be seen on table 3 that describe the value of section modulus of T1 until T29, on which the fluctuation of the section modulus value as can be seen in fig. 7 in accordance with the stiffeners in sequence of position, affected by the position of the stiffener on the inner and outer of the side and bottom of structure, including longitudinally center of the structure, respectively.
The formula of section modulus defines as follows:

$$Z = \frac{I}{y} \qquad (1)$$

Sample formula of moment inertia (second moment)

$$I = \frac{1}{12} bh^3 \qquad (2)$$

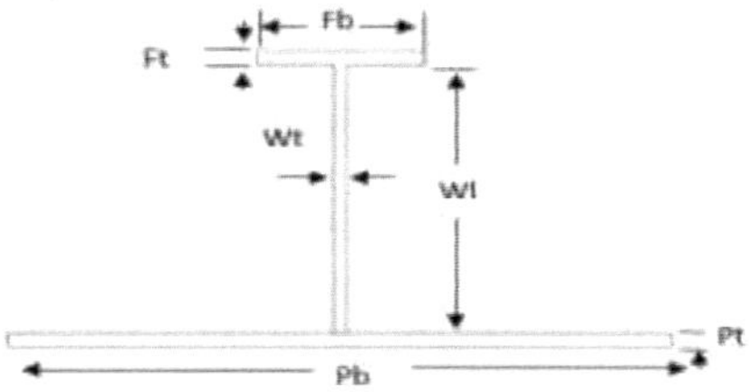

Fig. 5. T stiffener profile

Table 3. Value of section modulus of 29 convetional stiffener profiles

T	Section Modulus
T1	3402081.662
T2	3760164.146
T3	4120588.964
T4	4717603.407
T5	5495753.88
T6	5578789.863
T7	5711393.036
T8	6135085.767
T9	7747515.507
T10	3316365.262
T11	3421743.316
T12	4132737.746
T13	4251736.567
T14	4365117.581
T15	4979208.207

T	Section Modulus
T1	3402081.662
T2	3760164.146
T3	4120588.964
T4	4717603.407
T5	5495753.88
T6	5578789.863
T7	5711393.036
T8	6135085.767
T9	7747515.507
T10	3316365.262
T11	3421743.316
T12	4132737.746
T13	4251736.567
T14	4365117.581
T15	4979208.207

The difference among position of the stiffeners affects the geometrical dimension of each stiffener, and furthermore the section modulus. Fig.6 - Fig. 7 show that the smallest value of section modulus is belongs to the stiffener of T10, and the highest one is the stiffener of T29. The random pattern of section modulus value among 29 stiffeners is affected by the position of the stiffener, which has strong corellation with load working on the ship during the operational condition. Accordingly, the dimensional variables of the stiffeners e.g. flange breadth, flange thickness, web length, plate thickness varied for each positions. A mapping then conducted to elaborate the effect of the dimensional variable alteration toward the value of the section modulus and the ultimate strength of each stiffener. It started from the grouping of the flange thickness into 2 categories in sequence for its thickness i.e. 15 mm and 18 mm. Afterwards, each of the group is re-group according to the other dimensional variabel and sort from the largest into the smalles value. This conducted repeatedly for all geometrical dimensional variable of 29 stiffeners.

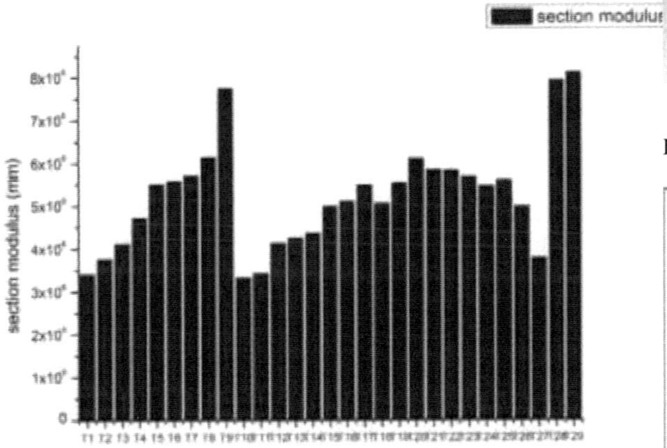

Fig. 6 Section modulus of 29 convetional T stiffners profile

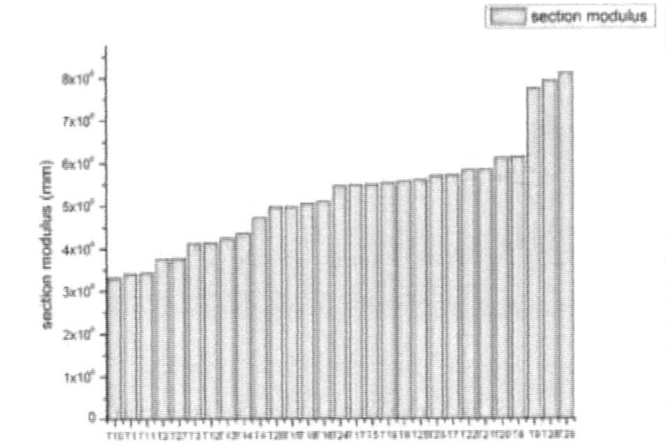

Fig. 7 Sorted of section modulus of 29 convetional T stiffeners profile

5. Load-displacement curve

Sagging and hogging phenomenon during the operation of the ship on ocean causes various kind of load on the ship structure, one among them is the compressive load. All of the structures are being given by such load, including the stiffener. The capacity of the stiffener on resisting the load is best describe in load-displacement curve. The actual compressive load working on the stiffener is considered on the elastic region of stress-strain curve, for the large value of section modulus in terms of cross sectional area i.e. an elastic deformation in relatively small longitudinal displacement after being given a massive load longitudinally, as shown in fig 8. The contour of stress distribution on fig. 8 describe the total deformation and it is related to the boundary condition, on which the fixed support is only in one side of the cross sectional area, and the other side become the area of the attached compressive load.

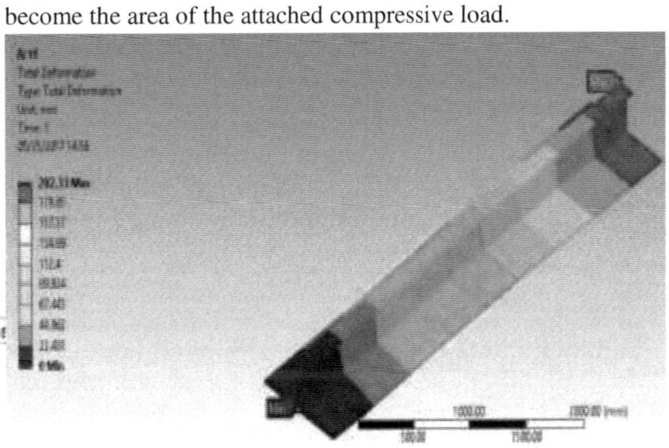

Fig. 8 Compressive load contour stress distribution

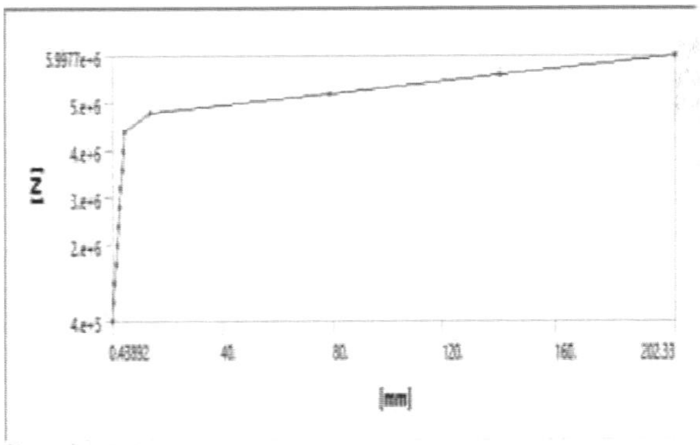

Fig. 9 Load-displacement curve

Fig 9 describes the ultimate strength characteristics of stiffeners, on which the given compressive load is covering the elastic-plastic area. The significant raise on the compressive load value working on the cross sectional area of the stiffener and the small amount of displacement on the elastic area i.e. the smallest is 0.43892mm and the largest one is 202.33mm. This is in line with the characteristic of compressive load in longitudinal direction of stiffeners i.e. small displacement for significant raise of load is affected by the geometrical dimension variable, cross sectional area and the section modulus. All of the 29 stiffener profile with variations of geometrical dimension was simulated using ANSYS FEM, in order to get the characteristics of the ultimate strength and its correlation to section modulus, as shown in fig 10. The load-displacement curve is related to the elastic and plastic area of deformation. Accordingly, it can be seen that the typical stiffener of T10 has the largest plastic deformation area and stiffener of T29 profile has only the elastic deformation area, in the other word stiffener of T 29 has small deformation area.

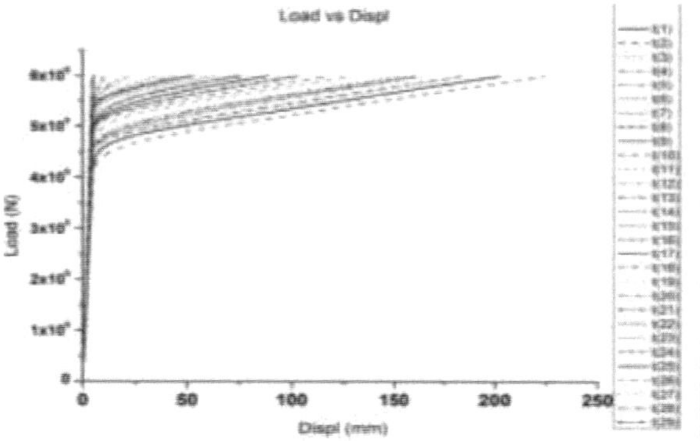

Fig. 10 Load-displacement of 29 stiffener profiles

After being given a continue raises of compressive load, the displacement is altered. The elastic area become the first region, furthermore in line with the raise of the compressive load, the stiffeners with the small value of section modulus e.g. T10, T1, T11 and so on, as shown in fig. 7, each of the stiffener has specific value of compressive load when entering the plastic region with increasing amount of displacement. Meanwhile, the stiffeners with large amount of section modulus e.g. T29, T28, T9 and so on, as show in fig.7, keep in the elastic region with raises of compressive load value, and on particular value of compressive load each of the stiffeners started to enter the plastic regions. However, there are stiffeners which keep on the elastic regions until the highest compressive load is being given e.g. T29 and T28, as shown in fig. 10. The role of section modulus is on the elastic-plastic region, the higher value of section modulus, the higher value of compressive load required to shift from elastic region into plastic regions, this is also related to the ability of the stiffener in adsorbing the energy of the compressive load.

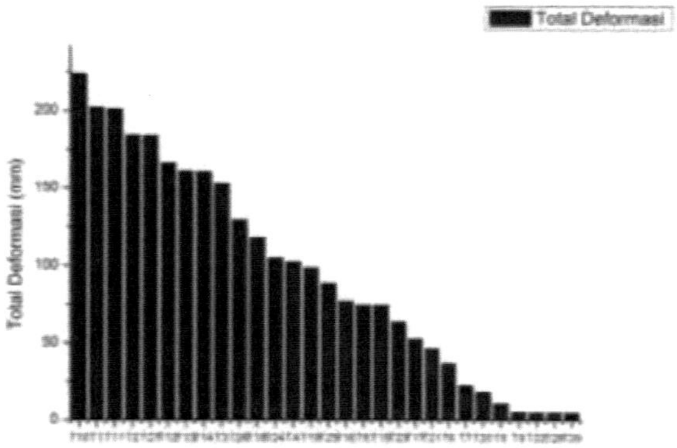

Fig. 11 Total deformation

Another role of section modulus is toward the total deformation, stiffener of T10 which has the lowest section modulus value also has the largest total deformation caused by the compressive load, as shown in fig. 11. In line with this, the stiffener T29, T28, T22 and T9 are the top four of the highest value of section modulus, having relatively similar the smallest value of total deformation. A mapping toward section modulus and total deformation is shown in fig. 12. The general tendency of 29 stiffeners for correlation between section modulus and total deformation is the larger value of section modulus of the stiffeners, smaller value of total deformation resulted. However, the raising value of section modulus has no linear pattern toward in ultimate strength as shown in fig 12, for instance the section modulus of stiffener T12 is larger than T3, but the ultimate strength of stiffener T12 is lower compared to the stiffener of T3. This phenomenon happened repeatedly on several different stiffeners e.g. T4 toward T26, T15, and T18; T16 toward T24; T17 toward T15, and T19; T6 toward T25, and T23; T6 toward T25, and T23; T22 toward T21, T20, T8, and T9. The pattern of the 29 stiffeners is related to the positioning of the stiffeners on the mid-ship section structure and the difference of geometrical dimension on each stiffener which refer to the international association classification society.

5. Geometrical effect of 29 stiffeners toward
6. Section modulus, ultimate strength and mass

The geometrical characteristics of each of the 29 stiffeners is mapped toward the ultimate strength in order to elaborate more regarding the factor that play important role in raise value that

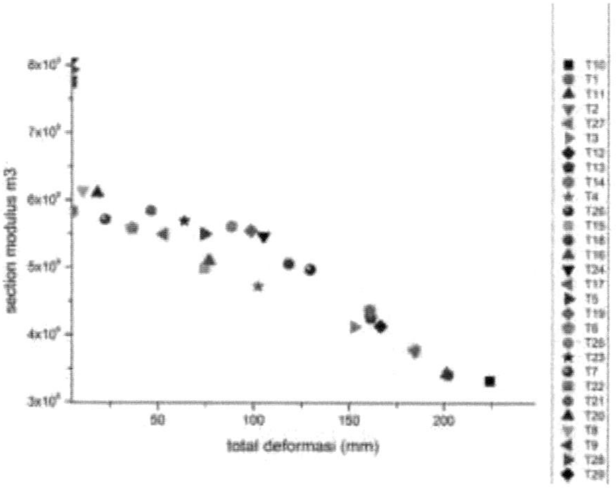

has no linear correlation toward the increasing ultimate strength. By finding the patterns of alteration of variable dimensional characteristics affecting the section modulus and ultimate strength, so that an evaluation for the efficiency of area improvement in classification society can be conducted. The mapping started with dimensional variable e.g. plate breadth, plate thickness, web length, web thickness, flange breadth and flange thickness. It is sorted in accordance with the smallest value of section modulus into the highest one, as shown in table 4. The T10 and T1 have similar value in all dimensional variables, but a raise in plate thickness gives increasing value of section modulus and ultimate strength. Other cases similarly happen on several stiffeners e.g. T13 and T14; T15 and T17; T6 and T7; T21and T20, T9 and T28. The raises value of plate thickness in line with the raises of section modulus and the ultimate strength. This is also valid for several stiffeners in terms of flange thickness e.g. T10 and T11; T2 and T27, and also web length T1 and T2; T14 and T15; T27 and T12; T12, T13 and T26; T18 and T 19; T7 and T8.

Table 4 variable dimensional variable of stiffeners, section modulus and ultimate strength

Regarding the effect of mass toward section modulus, it is generally found that a raise in section modulus value is linear with the increasing amount of mass of the stiffeners. However, there are several stiffeners having different pattern accordingly i.e. if the sort is in accordance with the lowest into the highest value of section modulus. A fluctuation of value of mass is happen for several stiffeners e.g. T12, T26, T18, T24, T25, T5, and T21. This is also valid for sorting conducted in terms of starting from the lowest value of mass to the highest one, the stiffeners having fluctuation of the section modulus belongs to

T3, T15, T4, T24, T16, T5, T17, T6, T7, dan T22. This shows that mass has no significant effect toward section modulus. Accordingly, the fluctuation on the mentioned stiffeners has correlation toward the geometry dimension parameter i.e. the T15 has strong correlation with the web length of T15 and T17, and plate thickness of T14 and T15. It seems that the significant raises of web length give contribution to increasing amount of the mass, as shown in table 5. However, the value of section modulus decrease when the web length is smaller compared to the raised stiffeners. In addition the section modulus decreases with the increasing mass when the flange breadth, plate thickness and web thickness, although the value of web length is larger. In accordance with the mass correlation toward the ultimate strength, generally the mass will give strong contribution in ultimate strength performance. Although, a finding for two stiffeners that has different pattern is caused by the in optimal design of the stiffener itself. The raises of web length and plate thickness including the effect toward ultimate strength and section modulus in percentage, elaborated as follows.

Table 5 Effect of web length and plate thickness toward percentage raise of ultimate strength and section modulus

T	Flange Breadth	Flange Thickness	Web Length	Web Thickness	Plate Breadth	Plate Thickness	Section Modulus	MASS	Deformasi	peningkatan % def	peningkatan % section modulus
T1	150	15	300	12	850	15.5	3402081.662	642.188875	202.33	8.86	9.52
T2	150	15	330	12	850	15.5	3760164.146	654.340675	184.41		
T4	150	18	380	12	850	16.5	4717603.407	718.475175	102.33	27.25	14.16
T5	150	18	440	12	850	16.5	5495753.88	742.778775	74.441		
T7	200	18	420	12	850	17.5	5711393.036	793.748825	22.273	52.46	6.91
T8	200	18	450	12	850	17.5	6135085.767	805.900625	10.588		
T12	150	18	360	12	850	15	4132737.746	667.33635	166.37	3.04	2.80
T13	150	18	370	12	850	15	4251736.567	671.38695	161.32		
T18	150	18	430	12	840	15.5	5051201.875	704.8044	118.12	16.48	3.85
T19	150	18	470	12	840	15.5	5541659.675	721.0068	98.659		
T27	150	18	330	12	850	15	3776638.112	655.18455	184.06	29.75	24.00
T26	150	18	430	12	850	15	4969323.921	695.69055	129.31		

T	Flange Breadth	Flange Thickness	Web Length	Web Thickness	Plate Breadth	Plate Thickness	Section Modulus	MASS	Deformasi	peningkatan % def	peningkatan % section modulus
T13	150	18	370	12	850	15	4251736.567	671.39	161.32	11.92040665	2.597433207
T14	150	18	370	12	850	15.5	4365117.581	685.73	142.09		
T15	150	18	420	12	850	15.5	4979208.207	705.99	116.49	54.67507941	9.264751358
T17	150	18	420	12	850	17.5	5487622.817	763.37	52.80		
T21	150	18	450	12	840	17.5	5838213.557	769.61	46.26	60.70131442	4.331171223
T20	150	18	450	12	840	18.5	6102524.335	797.97	18.18		
T24	175	18	450	12	850	15.5	5461632.162	733.33	105.04	15.63309216	2.542261703
T25	175	18	450	12	850	16	5604103.13	747.67	88.62		

7. Concluding Remarks

The main objective of this works is to numerically elaborate the geometrical effect of stiffener toward the ultimate strength of the conventional T stiffener profile. In accordance with this particular circumstance a series of non-linier finite element method computations were carried out on stiffeners. Geometrical dimension of variable of the stiffeners, section modulus, mass and ultimate strength become the variables, respectively.

Based on the results obtained from the present study, the conclusions can be described as follows:

- The alteration of section modulus value has no significant effect toward stiffeners strength i.e. the raise of section modulus was not followed by the increasing strength of the stiffener. The geometrical dimensions of some variables affect the strength of the stiffeners partially.
- Geometrical dimension strongly affect the strength of the stiffeners under longitudinal compressive load. The geometrical dimension variable that has strong contribution to the strength of the stiffeners is the plate thickness and web thickness, thicker the plate and web the stronger is the stiffener.
- The mass alteration of the stiffener affects the raise of the strength. Larger the mass higher value of ultimate strength in terms of longitudinal compressive load.
- The raise of plate thickness and web thickness has positive effect in toward ultimate strength and section modulus. A fluctuation of percentage raise of the ultimate strength and section modulus happen with the increasing value of thickness in web and plate.

8. References

9. [1] Hughes O. F, Paik J. K. Ship Structural Analysis and Design. The Society of Naval Achitect and Marine Engineer. 2010.
10. [2] Leheta H. B, Badran S. F, Elhanafi AS. Ship structural integrity using new stiffened plate. Thin-walled Structure 94(2015) 545-561.
11. [3] Amlashi H. K.K., Moan T. Ultimate strength analysis of a bulk carrier hull girder under alternate hold loading condition – A case study Part 1: Nonlinear finite element modeling and ultimate hull girder capacity. Marine Structures 21 (2008) 327–352.
12. [4] CommitT III.1. Ultimate strength. 19th International Ship and Offshore Structures Congress Volume I. 7–10 September 2015 Cascais, Portugal.
13. [5] Xu M. C, Soares C. G. Comparisons of calculations with experiments on the ultimate strength of wide stiffened panels. Marine Structures 31 (2013) 82–101
14. [6] Xu M.C., Soares C. G. Assessment of residual ultimate strength for wide dented stiffened panels subjected to compressive loads. Engineering Structures 49 (2013) 316–328.
15. [7] Xu M. C, Soares C. G. Effect of a central dent on the ultimate strength of narrow stiffened panels under axial compression. International Journal of Mechanical Sciences 100 (2015) 68–79.
16. [8] Ao L, De-Yu Wang D.Y. Ultimate torsional strength of cracked stiffened box girders with a large deck opening. International Journal of Naval Architecture and Ocean Engineering 8 (2016).
17. [9] Shi G., Wang D.Y. Residual ultimate strength of cracked box girders under torsional loading. Ocean Engineering 43 (2012) 102–112.
18. [10] Paik J.K., Seo J.K. Nonlinear finite element method models for ultimate strength analysis of sTl stiffened-plate structures under combined biaxial compression and lateral pressure actions—Part II: Stiffened panels. Thin-Walled Structures 47 (2009) 998–1007.
19. [11] Seo J.K., Paik J.K. Nonlinear finite element method models for ultimate strength analysis of sTl stiffened-plate structures under combined biaxial compression and lateral pressure actions—Part I: Plate elements. Thin-Walled Structures 47 (2009) 1008–1017
20. [12]Paik J.K.. Residual ultimate strength of sTl plates with longitudinal cracks under axial compression—Nonlinear finite element method investigations. Ocean Engineering 36 (2009) 266–276.
21. [13] Do H. C , JIANG W, JIN J, CHEN X. Ultimate limit state assessment of stiffened panel structures for very large ore carrier via nonlinear finite element method. International Journal of Recent advances in Mechanical Engineering (IJMECH) Vol.2, No.2, May 2013.
22. [14] Badran S. F., Saddek A. B, Leheta H. W. Ultimate strength of Y and T stiffeners subjected to

lateral loads with three different levels of initial imperfection.

Dynamic Analysis of PLTGL-SB by Simulation and Experimental Techniques

I K A P Utama, Department of Naval Architecture, Institut Teknologi Sepuluh Nopember, Surabaya, Indonesia
I S Arief, Department of Marine Engineering, Institut Teknologi Sepuluh Nopember, Surabaya, Indonesia
R Hantoro, Department of Engineering Physics, Institut Teknologi Sepuluh Nopember, Surabaya, Indonesia
J Prananda, Department of Marine Engineering, Institut Teknologi Sepuluh Nopember, Surabaya, Indonesi

SUMMARY

Wave Energy Converters (WECs) plant based on pendulum system is one type of the WECs that has easy ways of working. In this task, the WECs based on pendulum system had been tested using simulation CFD (Computational Fluid Dynamic) and experimental approach. The aim of this research is to analyze the geometry of pontoon that exacts to trimaran that can produce a huge amplitude of pendulum oscillation. The analyzed aspect is the combination of outrigger length, outrigger height, pendulum rod length, pendulum mass, and wave periods which have a maximum amplitude of pendulum oscillation. The results show that the best combination is pontoon thus has outrigger length of 413 mm, outrigger height of 40 mm, pendulum rod length of 106.7 mm, pendulum mass of 20, and wave periods of 0.8 s which have a maximum amplitude of pendulum oscillation as big as 60 degrees. The pontoon response (RAO) roll, heave, and pitch on draught 1/2 or 2/3. The frequency around 1.25 Hz - 1.5 Hz decreased.

NOMENCLATURE

[symbol]	Definition [(unit)]
ρ_n	Density of the object (kg/m^3)
$m_{tot} = \sum \rho_n V_n$	Volume of the object (m^3)
$B_{tot} = \sum m_n g$	Gravity (m/s^2)
B	Weight of object (N)
I_O	Moment Of Inertia (Kg.m^2)
$\omega = \frac{\sqrt{2gh}}{l}$	Angular Velocity (rad/s)
$P = \tau\omega$	Power (watt)
$P = \tau\omega$	Torque (N.m)
$h = h - h\cos\theta$	Height (m)
l	Length (m)

1. INTRODUCTION

Energy is number one global primary demand. Data from BP (British Petroleum), the growth of economy following requires more energy, the increase of energy demand by falls in energy intensity (energy used per unit GDP) [1]. Every country must think the solution to provide the energy demand by the increase of its GDP.

A green economy makes as the results to make enhanced human well-being and social equity defined by UNEP (United Nations Environmental Programme) [2]. International Energy Agency (IEA) defined Renewable Energy generation to support the green economy for devoid greenhouse pollution and provides energy security [3]. Renewable Energy is one solution to make better life for society

One source of renewable energy that has the potential to be converted into electrical energy is derived from the sea its called Wave Energy Converters (WECs). WECs can be broadly classified into several types, namely tidal energy, wave energy, ocean thermal energy, ocean currents energy, and sea wind energy and Salt gradient [4]. The three most common types of marine energy are tidal energy, wave energy, and ocean thermal energy [4]. Sea waves as one of renewable energy are known as resources that have the potential to be exploited in a sustainable manner to be converted into electrical energy [5].

As in the world, electricity demand in Indonesia is also increasing along with the times. In addition, there are at least three major groups of national energy problems: low electrification rates, dependence on fossil energy sources, and low utilization of renewable energy [6]. Ministry of Energy and Minerals in Indonesia make policy in 2025, 23% of total energy must be produced by Renewable Energy [7]. To pursue future energy needs the policy to be done is to integrate renewable and nonrenewable energy, with a greater change in the composition of use in renewable energy types.

Indonesia itself is a maritime country with great wave energy potential. Our coastal line is no.2 in the world after Canada (95,181 km) [8]. Ocean wave energy is a renewable energy that is continuous and environmentally friendly. In addition, there is also a tool of ocean wave energy conversion into electrical energy developed by researchers from Indonesia that is using pendulum system or called as a wave powerhouse wave pendulum system (PLTGL-SB). The study of PLTGL of this pendulum system was first performed by Zamrisyaf, one of the PLN planning staffs in West Sumatra. The concept of a wave powerhouse of the pendulum system is simple, that is, by exploiting the fluctuation of the movement of sea water waves which will make the pendulum move like a bell, the movement is then transmitted into a rotary motion to move the dynamo then into electrical energy [9]. The power generated by the wave powerhouse of the pendulum system is much influenced by the dimensions of the pontoon, the length of the pendulum, and the mass of pendulum [10]. The parameters affecting the magnitude of power are the amount of torque associated with the amplitude and the frequency of the resulting vibration [11].

In earlier studies, the previous pontoons were rectangles shape made by Zamrisyaf. The authors will analyze the best dimension of pontoon using CFD (Computational Fluid Dynamic) explored the power generated by PLTGL-SB system with cylindrical pontoon which has symmetrical outrigger on both sides. The authors also examine the relationship of mass, length of the pendulum arm, the length of the outrigger sides, and the outrigger heights from the base to the torque generated by the pendulum. The research will be done through experiments in the laboratory by using a prototype that has a scale of 1:10 to the real dimension.

2. Methods

2.1 Literature Study

Literature studies include an understanding of the concept of the research to be undertaken from the review of supporting journals to the collection of data required to support this research. In this case it understands the working principles of the pendulum wave power system, the understanding of the flow of submerged objects, the understanding of hydrodynamics, the understanding of flow separation, the aspect ratio of turbine design, and the taking of data from the wave size created for use in the experiment up to the frequency and its equipment settings. This literature study also helps researchers to determine the dimensions of the tool to be made related to the stability and response of the movement of the object when it is given a wave.

2.2 Analytic Calculation

Once determined the dimensions of the pontoon prototype will be made, then the analytical calculation to support the data so that it can be known combination of sizes in accordance with the desired specification, the condition where the pontoon continues to shake but still in a stable positive condition. The dimensions of the self-made pontoon prototype are 32 cm long cylinders with 16.5 cm diameter. which has a symmetrical outrigger on both sides as long as 32 cm with a diameter of 4 cm. A pendulum made is a vertical pendulum. While the given wave direction is 90 degrees from the face of the pontoon.

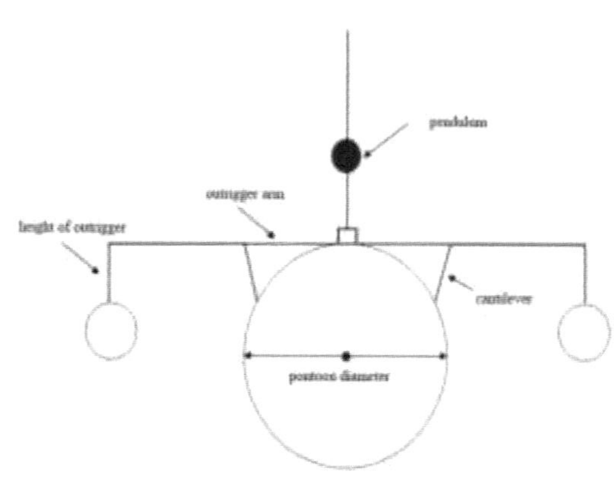

Figure 1: Front view Pontoon Design

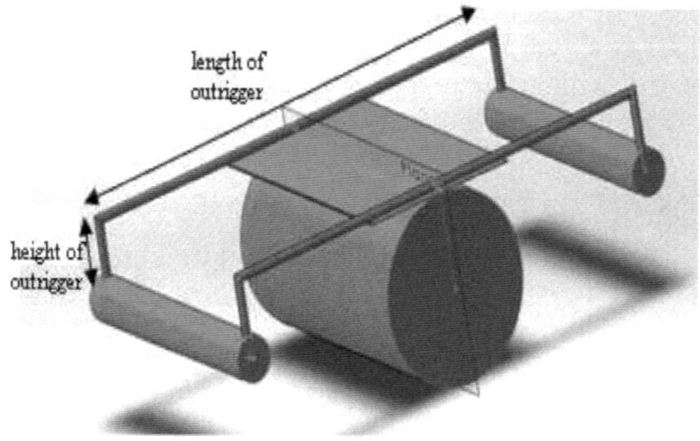

Figure 2: Pontoon 3D Design

Specification of the design of the pontoon prototype to be performed calculation analysis is as follows.

Large Cylinder Diameter: 16.5 cm

Large Cylindrical Thickness: 0.7 cm

Small Cylinder Diameter: 4 cm

Thickness of Small Cylinders: 0.2 cm

Large Cylinder Length: 32 cm

Small Cylinder Length: 32 cm

Mass Cylindrical Type: 1.39 gr / cm^3

Full: ½ and 2/3 diameter of pontoon

High of Outrigger: 1.5 cm and 4 cm

Plate length: 32 cm

Plate width: 17 cm

Plate Density: 1.2 gr / cm^3

Once determined the main size of the pontoon dimension along with the materials used, then calculated parameters that will determine the stability of the prototype made. The parameters are COB, COG, and metacenter. The first thing to be calculated is to determine the mass of each component so that it can be known the total mass of the whole system. The mass calculation can be done using equation 2.1 with ρ is the density of the object and V is the volume of the object. After the mass of the object is found, it is necessary to look for the weight of the object as equation 2.2.

$$m_{tot} = \sum \rho_n V_n \qquad (2.\ 1)$$

$$B_{tot} = \sum m_n g \qquad (2.\ 2)$$

When the weight of each component and total weight has been found, the center of gravity (COG) can be searched using equation 2.3 where B is the weight of the object and l is the center point of each section in terms of the abscissa axis.

$$COG = \frac{B_1 l_1 + B_2 l_2 + B_n l_n}{B_{tot}} \qquad (2.\ 3)$$

Meanwhile, to find the value of the center of buoyancy (COB), will be more easily searched because we have determined from the beginning the size of the submerged pontoon. COB is likened to COG from the submerged parts, so the search will be the same as COG with a record of just the submerged pontoon that is used as a review. Thus, COB can be searched using equation 2.4.

$$\mathrm{COB} = \frac{B_{submerged} \,.\, l_{submerged}}{B_{submerged}} \qquad (2.4)$$

The next step is to calculate the metacenter of the pontoon. Metacenter itself can be searched using equation 2.5 with I_0 is the moment of system inertia, Vt is the volume of the pontoon submerged, and G0B0 is the distance between COG and COB.

$$M = \frac{I_0}{V_t} - G_0 B_0 \qquad (2.\ 4)$$

On a floating object, the magnitude of the Archimedes style will be the same as the force of gravity. So from the principle, we can know how big the submerged so that we can find out how big ballast needed for the pontoon can reach laden pre-determined. Then to determine the ballast, can be searched as equation 2.6 to 2.8 with ρf is the fluid type mass, Vt is the volume of the submerged pontoon, ms is the total mass of the system so that the pontoon can be submerged in accordance with the specified load, and g is the acceleration of earth gravity.

$$Fa = m_s g \qquad (2.\ 5)$$

$$\rho_f V_t g = m_s g \qquad (2.\ 6)$$

$$m_s = \rho_f V_t \qquad (2.\ 7)$$

If the mass of the system has been obtained, then the ballast can be found by reducing the mass of the system with the total mass of the pontoon previously calculated using equation 2.1.

2.3 Simulation Using CFD

In finding the best geometry of pontoon, the authors used CFD (Computational Fluid Dynamic). Stage modeling pontoons things that must be input, including:

- Water depth, on the geometry
- Mass fluid type (water), on the geometry
- Water size on the x and y-axes, in the geometry section
- Point COG (Center of Gravity), at the point mass
- Grating radius, at the point mass.

The depth of waters inputted in accordance with the depth of the tank where the experiments tool is done, as deep as 2 meters. Water size on the x-axis is 50 m and y-axis of 3 m, according to the tank where the experiment is done.

Figure 3: PLTGL-SB in CFD software

2.4 Determination of Prototype

After the analytical calculation and CFD simulation, it can be determined the feasibility of the prototype related with specifications that have been in accordance with the purpose of research. If at the previous calculation the metacenter point is at the top, followed by COG, then COB, then the pontoon has stabilized positively so that a predetermined specification can be made.

In this experiment, there were 48 variations with 3 variations of wavelength periods, 2 variations of the outer sleeve length, 2 variations of the outrigger that can be represented by water-laden, 2 variations of pendulum arm length, and 2 pendulum mass variations. The wave height used in the data retrieval process is 1.5 cm for all variations.

2.5 Prototype Fabrication

After the analytical calculation and determination of the dimensions of the pontoon, the prototype can be fabricated in accordance with the design specifications. Figures 4 and 5 show the pontoon prototype that has been created.

Figure 4: Pontoon with ballast 2.3 kg

Figure 5: Pontoon with ballast 1.2 kg

2.6 Data retrieval

In this experimental study, the data were collected in the towing tank of Hydrodynamics Laboratory of Marine Faculty Technology-ITS. Which has a length of 50 m, width 3 m, and a depth is 2 m. The data taken is the video movement of the pontoon and pendulum to know the deviation, as well as the amplitude of the roll and heave of the pontoons using sensors that have been available in the laboratory.

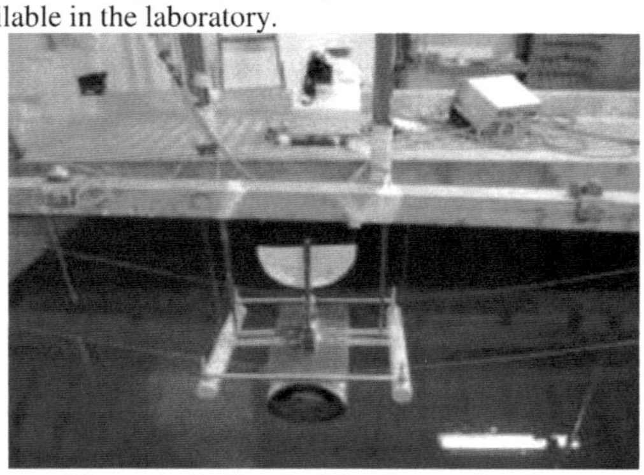

Figure 6: Setting tool for data retrieval

2.7 Data Analysis and Paper Making

The data obtained from the experiments are the roll and heave drift of the pontoon and also the videotape of the pendulum data. The first thing to do is to process the video so that the data of pendulum deviation is quantitative. After the data obtained pendulum deviation and also pontoons, then done through data so that obtained graphs and tables that can represent the data. In addition, it is necessary to calculate the power generated by each condition of the experiments performed. Before sought power, the torque value (τ) must be known. The torque can be known using equation 2.9 where m is the pendulum mass, g is the acceleration of gravity, l is the length of the pendulum arm, and θ is the junction at the pendulum oscillation.

$$\tau = mgl\sin\theta \qquad (2.8)$$

In addition to torque, the angular velocity of the pendulum oscillations should also be known. The angular velocity of the pendulum itself can be determined using the equation 2.10 with g is the acceleration of gravity, l is the length of the pendulum arm, and h is the difference of the pendulum height when in the θ junction which can be searched using equation 2.11. After knowing torque and angular velocity, power can be searched using equation 2.12.

$$\omega = \frac{\sqrt{2gh}}{l} \qquad (2.9)$$

$$h = h - h\cos\theta \qquad (2.10)$$

$$P = \tau\omega \qquad (2.11)$$

After all the data is processed, then the preparation of the making paper which is the final stage of this research project work and do as delivery of results and analysis of the implementation of research programs. The paper refers to the previous stages and describes the entire process of activity using a particular systematic.

3. RESULTS AND DISCUSSION

3.1 Pontoon Dimensions

The tested pontoon is a prototype with a ratio of 1:10. Prototype pontoons are made of cylindrical shape with outrigger symmetrical on both sides with a length of 320 mm. The diameter of the pontoon is 165 mm, while the outrigger has a diameter of 40 mm. For the pendulum used consisted of 2 variations of mass, ie 20 grams and 40 grams with the variation of arm length 106.7 mm and 165 mm. For more details, the shape of the prototype and pendulum can be seen in the picture below.

Figure 7: Dimension of the pontoon with ballast 1.2 kg

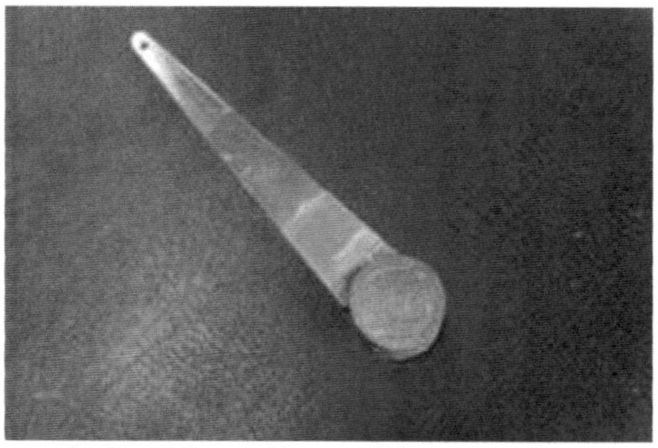

Figure 8: Pendulum with mass 20 gram and arm length 106.7 mm

3.2 Mass Calculation, Volume, and Stability of Pontoons

Once the geometry of the pontoon is determined, it is necessary to calculate the stability of the pontoon when exposed to the excitation force of the wave. The thing that most affects the stability of an object when immersed in water is the number of submerged parts of the object. The height of the part of the object dyed in the water is called as laden (t). If the load is known, then the center of buoyancy (COB) will also be known. To know all the data, it is necessary to calculate the mass and volume of the pontoon to be investigated.

Table. 1 Weight of Pontoon and Pendulum Construction

No.	Part of Pontoon	Volume (cm^3)	Density (gr/cm^3)	Mass (gr)
1.	Pontoon	1076.58	1.39	1496.45
2.	Outrigger	147.95	1.39	205.66
3.	Cover of the pontoon	213.72	1.20	256.46
4.	Cover of the outrigger	25.12	1.20	30.14
5.	Outrigger arm	7.61	2.71	57
6.	Outrigger arm 2	6.33	2.71	47.8
7.	Height of outrigger	15.2	2.71	70.68
8.	Pendulum construction	-	-	80.00

Based on the calculation, then obtained the total mass of the pontoon along with construction outrigger and pendulum that is equal to 2.12 kg. The selected self is 0.5 and 0.67 diameter of the pontoon. Through analytical calculations, it can be obtained that the location of COB, COG, and metacenter point in accordance with the theory of stability so that the pontoon can return to its original position (stable positive).

Fig. 8. Position points for positive stable objects (K = keel, B = COB, G = COG, and M = metacenter)

According to calculations, the mass of the system that is equal to 2.12 kg. However, with the desired weight of 2.12 kgs is not reached, the ballast is required. Data on point COG, COB, metacenter, and also ballast can be seen in table 2.

Table. 2 Pontoons Stability

Draught	COG	COB	Metacenter	Ballast
	(cm)	(cm)	(cm)	(kg)
0.5 D	8.67	7.49	12.17	1.2
0.67 D	9.26	7.68	11.78	2.3

3.3 Response of Pontoon with Variation of Water Draft using CFD

The graph below shows the movement response (RAO) roll, heave, and pitch on draught 1/2 or 2/3 after passing the frequency around 1.25 Hz - 1.5 Hz decreased, meaning that if the pontoon is placed on the frequency increase or small period then the movement response will decrease.

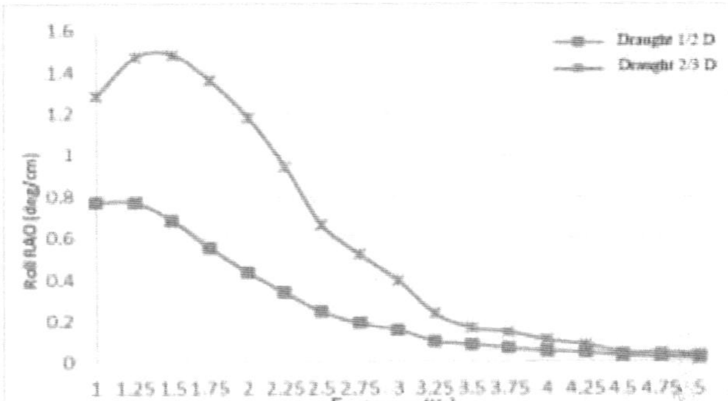

Figure 9: Roll RAO with Variation of Water Draft

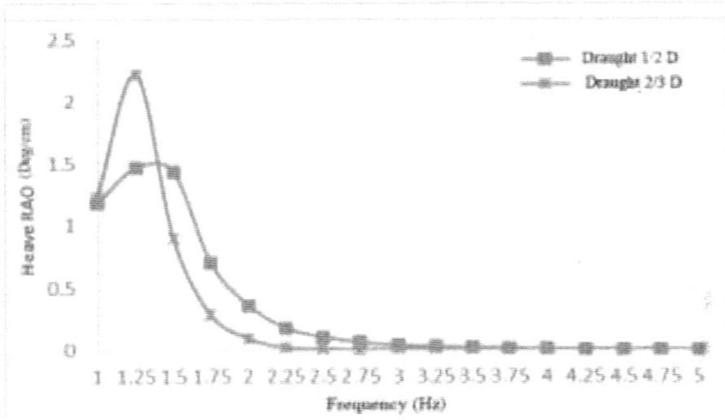

Figure 10: Heave RAO with Variation of Water Draft

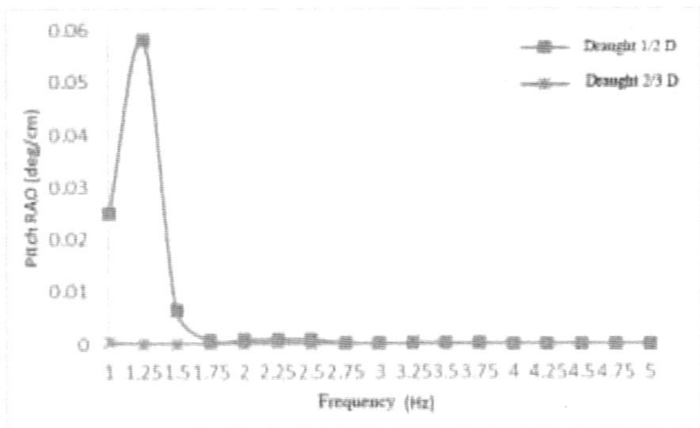

Figure 11: Pitch RAO with Variation of Water Draft

3.4 Experimental Results of Ponton and Pendulum Deviation

The number of variations tested was 48, with 3 variations of wavelength period, 2 variations of the outrigger sides, 2 variations of outrigger height, 2 pendulum arm variations, and 2 pendulum mass variations.

In general, the oscillation characteristics of the experimental results are divided into 4 major sections. The first characteristic is that there is no oscillation at all. This phenomenon occurs in variations that are given a coming wave with a period of 0.4 seconds.

The second characteristic is when the pendulum oscillates with a smaller amplitude equal to 20 degrees. This phenomenon occurs in several variations of the pontoon given the wave style with a period of 0.6 or 0.8 seconds.

The third characteristic is that the pendulum vibrates back and forth, not oscillating as expected by the researcher. This phenomenon occurs in 5 variations given the wave style with a period of 0.6 seconds (variation 17, variation 26, variation 32, variation 38, variation 47).

The fourth characteristic is when the amplitude of the pendulum oscillation is of great value, ie more than 20 degrees. This fourth characteristic occurs in some variations that are given waves with periods of 0.6 and 0.8 seconds.

From the four characteristics mentioned above, the fourth type is eligible to be used as an ocean wave power plant because it produces a large amplitude of oscillation so that it can produce great power as well. From several variations on the incoming of the fourth characteristic, there are three variations that produce the largest amplitude and maximum amplitude, ie variations 12, 36, and 42. Figure 12 to 14 shows the result of pendulum oscillation along with the heaving value and rolling of pontoons on each of these variations.

The largest mean amplitude occurring at the time-tested was a pontoon of variation 12 with an average amplitude value of 41.82 degrees and a maximum amplitude of 60 degrees. While the average amplitude at 36 and 42 variations in a row is 36.52 degrees and 32.54 degrees respectively. Variation 12 produces the largest pendulum shrinkage value because it is the most appropriate combination to obtain large amplitude values.

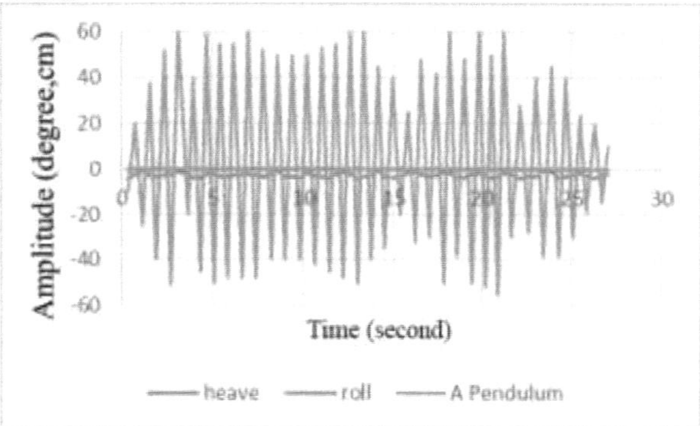

Figure 12: Motion of pontoon and pendulum for variation 12
(T=0.8 s; Pc=413 mm; Hc=40 mm; Lp=106.7 mm; Mp=20 gr)

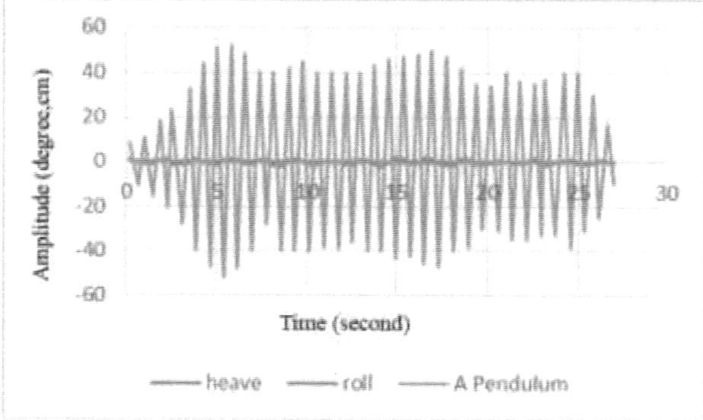

Figure 13: Motion of pontoon and pendulum for variation 36
(T=0.8 s; Pc=495 mm; Hc=15 mm; Lp=106.7 mm; Mp=20 gr)

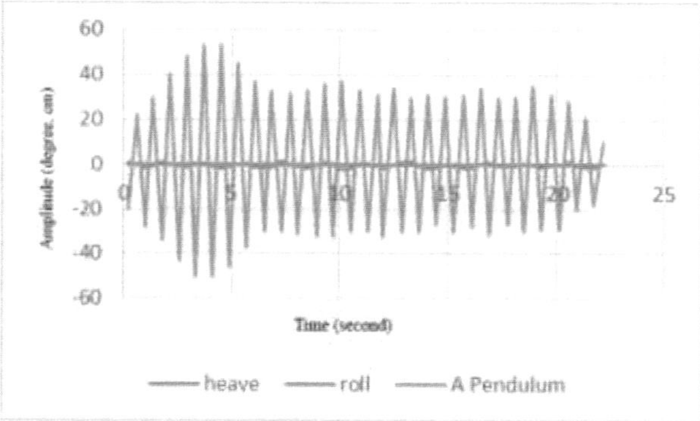

Figure 14: Motion of the pontoon and pendulum for variation 42
(T=0.8 s; Pc=413 mm; Hc=15 mm; Lp=106.7 mm; Mp=20 gr)

If the dimensional analysis is used, then when enlarged to the original scale variation 12 has a pontoon finger size of 1.65 m, outrigger fingers of 0.4 m, pontoon length and outrigger 3.2 m, 4.13m outrigger, outrigger 0.4 m, pendulum arm 1.067 m, and pendulum mass of 20 kg.

3.5 Effect of Variables on Pendulum Oscillation

After 48 experiments, the effect of the related variables on the pendulum oscillation. Data obtained that the greater the given wave period, the amplitude of the pendulum oscillation will also be greater. The greater the wave period does cause the rolling angle to become smaller, but not so with the rolling speed that occurs. If the rolling speed is large, the frequency of the pontoon vibrates also will be faster so that the pendulum will more easily oscillate.

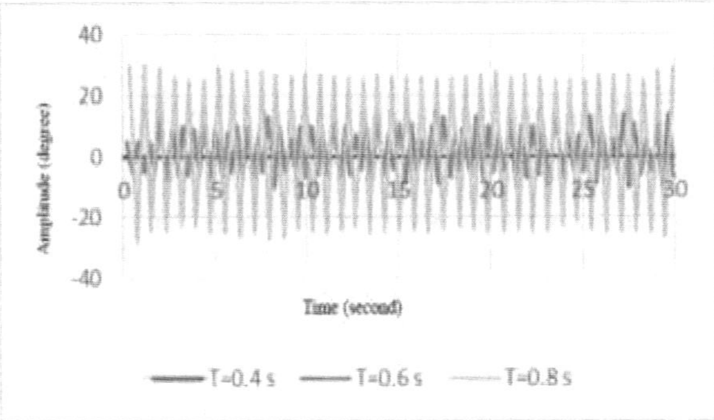

Figure 15: Pendulum oscillations with different wave periods

Based on Figure 15 the variation of 15 is when the pontoon has an outrigger with a 495 mm sleeve length from end to end, whereas, in variation 12, the length of the outrigger arm from end to end is 413 mm. Based on the experiment, we can see if the length of the outrigger arm is greater, the average amplitude generated by the pendulum oscillation will be smaller. At the time of variation 12 (outrigger sleeve = 413 mm), the average amplitude of pendulum oscillation is 41.82 degrees with a vibration frequency of 1.23 Hz. Whereas when a variation of 15 that has a 495 mm outrigger sleeve, the pendulum will oscillate with a maximum amplitude of 21.44 degrees and a frequency of 1.355 Hz. According to D. R. Berrett and C. B. Barrass (1999) by increasing the width of the ship (the length of the vessel is considered fixed), the vessel will be more stable [12]. This is what happens to the pontoon tested by the author. Seen in Figure 13 that the pontoon variation 15 with the larger outrigger sleeve length has a smaller rolling motion, then as a consequence, the pendulum oscillation will have a smaller amplitude as well.

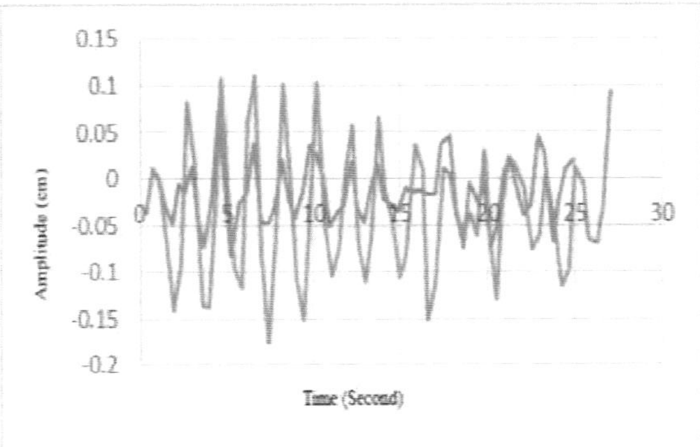

Figure 16: Rolling motion for variation for variation 12 and 15

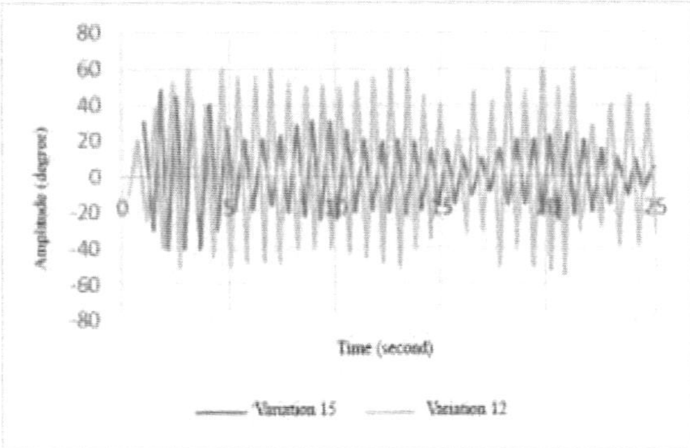

Figure 17: Pendulum oscillation with changes of the outrigger arm length

In this experiment, changes in the outrigger level will affect the magnitude of the pontoons that are submerged in water. The bigger the outrigger, the value of the pontoon will be smaller. The larger-laden pontoon (variation 27, laden as high as 106.7 mm) has a larger rolling amplitude compared to the 82.5 mm loaded pontoon. The larger rolling amplitude is why the variation 27 has a greater oscillation amplitude compared to the variation 6.

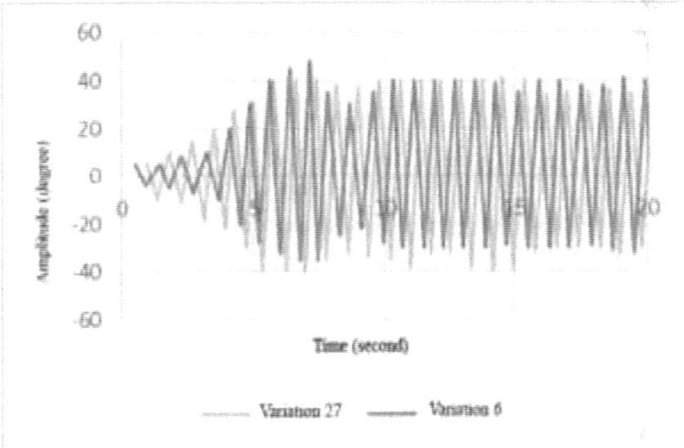

Figure 18: Pendulum oscillation with outrigger high change

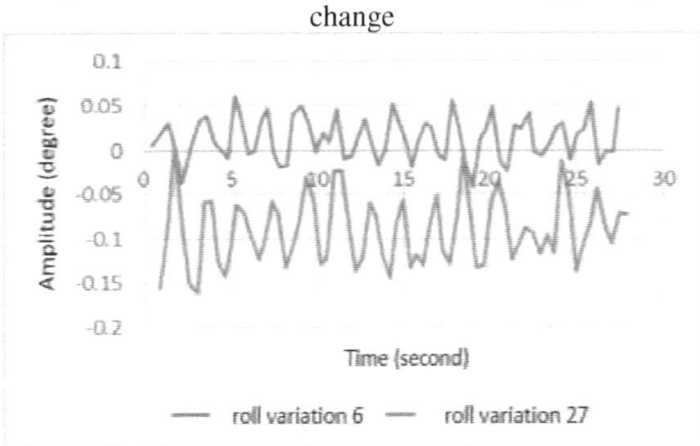

Figure 19: Motion rolling when variations 6 and 27

If we see again equation 2.9 it can be seen that the mass will be inversely proportional to the size of the pendulum. This also happens to experiments that have been done by the author. It can be seen figure 17 that the pontoon with a variation of 18 having a pendulum mass of 40 grams has a mean deviation smaller than when the pontoon tested is a variation of 21 with a pendulum mass of 20 grams. The pendulum amplitude at variation 18 is

16.61 degrees, while variation 21 has an average pendulum oscillation amplitude of 19.32 degrees. The frequency of the pendulum oscillation itself does not have a significant difference value so that in the calculation of power will not be too much effect.

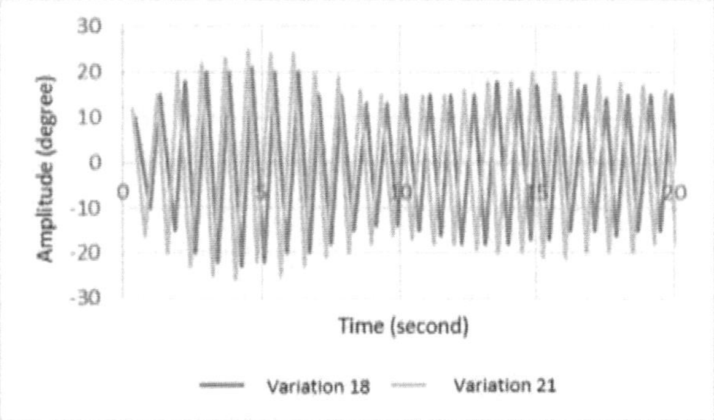

Figure 20: Pendulum oscillation with mass changes

we can see that the root of the pendulum arm will be proportional to the period. The angular deviation from the pendulum will be inversely proportional to the period. So it can be concluded that the shorter the pendulum arm the larger the deviation. This is why the pendulum pendants with longer pendulum sleeves have smaller deviations compared to pendulums with shorter arms as shown in Figure 20. Variation 30 was pontoon with pendulum sleeve of 106.7 mm with mean pendulum value of 38.01 degrees, while variation 33 has pendulum arm length of 165 mm with average pendulum deviation value of 31.87 degrees.

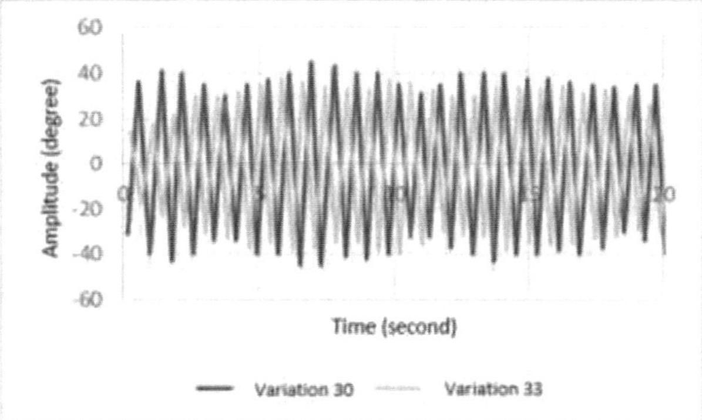

Figure 21: Pendulum oscillation with pendulum arm length change

The obtained the value of torque, omega, and average power of each variation as shown in Figure 21 and 22. It will be found that the average power and also the greatest angular velocity are obtained when variation 12 which at the time of the condition is obtained also the largest and highest mean deviation value among the other variations.

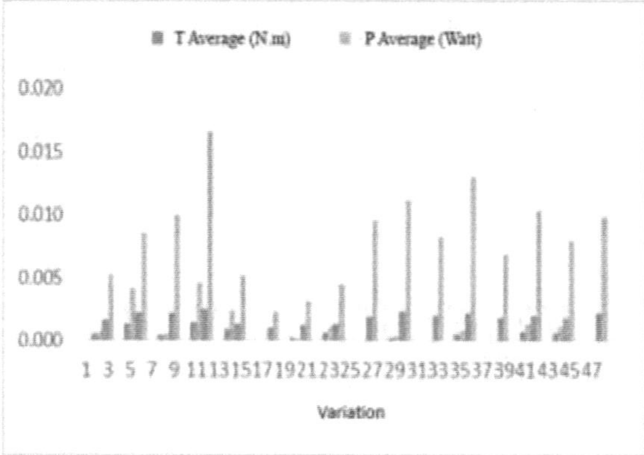

Figure 22: Torque, angular velocity, and average power on each variation

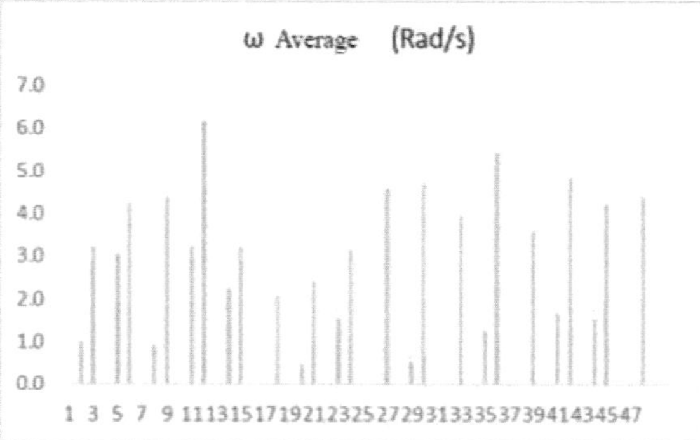

Figure 23: Average angular velocity on each variation

Based on analytical calculations, it was found that the power obtained for variation 12 is 0.017 watts with an average angular velocity of 6.173 m/s.

4. CONCLUSION

Referring to the results of data analysis and discussion in the previous chapter, it can be taken some conclusions about this experimental study as follows.

The amplitude of the pendulum oscillation decreases as the length of the outrigger arm, the outrigger, the pendulum mass, and the length of the pendulum arm have, but unlike the rise of the wave period, the pendulum oscillation amplitude increases as the parameter increases.

The variations that resulted in the maximum deviation value and the largest mean deviation were variation 12 with the wave period 0.8 s, the outer sleeve length 413 mm, the outrigger height 40 mm, the pendulum arm length of 106.7 mm, and the pendulum mass of 20 grams with a maximum deviation of 60 degrees and an average deviation of 41.82 degrees.

Some things that can be suggested from the results of this research for the continuation of its development. Further research can be a variation of the influence of the direction of the arrival of the wave of pontoon movement which will affect the pendulum oscillation.

Data retrieval on subsequent variations is done when the water conditions in the towing tank are completely quiet. In addition, the sensor also needs to be checked each turn of variation so that the data obtained can be really ideal and accurate.

5. ACKNOWLEDGMENTS

The authors gratefully thanks, acknowledgment funding support from ministry research and higher education in Indonesia through a research grant.

6. REFERENCES

BP British Petroleum (BP). BP Statistical Review of World Energy. 2017 edition.

United Nations Environment Programme. Towards a green economy: pathway to sustainable development and poverty eradication. http://www.unep.or/greeneconomy; 2011

IEA International Energy Agency (IEA). Tracking clean energy progress. IEA publications, Paris, Frances.

Indonesia Energy Outlook 2014, Badan Pengkajian dan Pengembangan Teknologi , 2014.

Pengenalan Teknologi Pemanfaatan Energi Angin, Tasikmalaya: Tim Lentera Angin Nusantara, 2014.

Tutorial Blade 3 Dimensi, Tasikmalaya: Tim Lentera Angin Nusantara, 2015.

Renstra ESDM Kementerian ESDM. 2016. "Program Strategis EBTKE dan Ketenagalistrikan". Jurnal Energi Media Komunikasi Kementerian Energi dan Sumber Daya Mineral. No. 2:10

"Coastal and Marine Ecosystems — Marine Jurisdictions: Coastline length".World Resources Institute.

H. Piggot, Windpower Workshop: Building Your Own Wind Turbine, Centre of Alternative Technology Publications.

S. Mehrangiz, Y. Emami, S. H. S. Sadigh and A. Etemadi, "Various Technologies for Producing Energy from Wave," *International Journal of Smart Grid and Clean Energy,* vol. 2, no. 2, pp. 289-294, 2013.

G. Iglesias, M. Alvarez and P. Garcia, "Wave Energy Converters," *EOLLS,* vol. 55, pp. 235-241, 201

B. Bryan and D.R. Derrett, Ship Stability For Masters And Mates, Oxford, UK: Butterworth-Heinemann, 2006.

7. AUTHORS BIOGRAPHY

I ketut Aria Pria Utama

I Ketut Aria Pria Utama is a professor of ship hydrodynamics at the Institute Teknologi Sepuluh Nopember (ITS). He obtained his PhD in 1999 from the University of Southampton. He is a fellow member of the Royal Institution of Naval Architects (RINA) since 2006 and member of the Indonesian Academy of Sciences (AIPI) since 2015.

His research interests include ship resistance and propulsion, seakeeping, computational fluid dynamics, renewable energy, and maritime engineering education.

Irfan Syarif Arief

Irfan Syarif Arief is a lecturer at the Department of Marine Engineering, Institut Teknologi Sepuluh Nopember (ITS), Surabaya. He obtained his Master Degree in 2003 focusing in computational fluid dynamics from Institut Teknologi Sepuluh Nopember Surabaya.

His research interests in computational fluid dynamics focusing on marine technology case includes ocean energy wave and tide energy, system propulsion, and rudder arrangement.

Ridho Hantoro

Dr. Ridho Hantoro is a senior lecturer at Department of Physics Engineering, Institut Teknologi Sepuluh Nopember (ITS), Surabaya. He obtain his doctorate Degree in 2001 focusing in marine renewable energy.

His research interests include energy management and numerical simulation in renewable energy and hydrokinetic turbine.

Juniarko Prananda

Juniarko Prananda is a associate lecturer at Department of Marine Engineering, Institut Teknologi Sepuluh Nopember (ITS), Surabaya. He obtain his master degree in 2014 focusing in control instrumentation and renewable energy.

His research interest include automation control system, wave and current energy.

Effectiveness of Nonlinear Roll Damping Formulations on Roll Response in Regular Waves

S. Rao TATA, South Asia Technical Support Office, Lloyds Register Asia, Singapore

SUMMARY

Method of multiple scales is used to obtain analytical solution of the nonlinear roll equation and to verify the effective and valid formulation of the nonlinear damping in regular waves. Roll motion is highly nonlinear in nature due to nonlinear damping and restoration moments. There are various forms that are used to represent nonlinear damping moment. In present study, the effect of form of nonlinear damping moment on estimation of roll response of a training ship is investigated. Damping coefficients are estimated from free roll decay curve and hydrodynamic theories. Frequency-response relations obtained with method of multiple scales are used to calculate roll response. Analytical results are compared with roll amplitudes obtained from forced roll experiments in regular beam waves. Results indicate that, damping moment representation has effect on estimation of maximum roll amplitude in regular beam waves.

Keywords: *nonlinear roll damping, free roll, multiple scales method, maximum roll amplitude*

1. Introduction

Among six motions of ship, roll has been in main attention over the years because of it being most critical motion leading to capsize. For most degrees of freedom, radiation damping due to waves is enough for estimation of ship motions at wide range of wave frequencies. For many vessels the natural frequency of roll motion falls within the range of frequencies at which wave energy dominates and small hydrodynamic damping also contribute to the large resonant response in roll motion. It is well known that roll damping is highly nonlinear in nature due to damping and restoring moments. In case of small angles, response can be well estimated by a linear equation. Nonlinearity becomes more important at larger angles.

Many authors have suggested various representations for nonlinear damping and restoring moment. Restoring moment is usually represented by various degree of odd order polynomial of roll angle. However there are many expressions defining nonlinear damping. Classical damping theory takes into account quadratic dependency from angular velocity. This representation is usually known as conventional linear plus quadratic velocity damping. Dalzal (1978) suggested various representations of damping moment[(1)]. He used a cubic dependency on roll velocity. This is referred as linear plus cubic velocity damping. Cardo.A et al. (1980) studied the effect of angle dependant damping on roll response in regular waves and showed nonlinearities attenuate maximum roll amplitude but capsize danger still remains[(2)]. They compared analytical maximum roll amplitudes calculated with 'angle dependent and cubic damping' with the existing roll experiments in literature (Wright J.G. H et al. 1979) [(3)]. They found good agreement with experiments. Cardo Antonio et al. (1982) gave generalised damping models, taking into dependency of roll angle on damping[(4)]. Haddara (1984) studied the effect of damping moment on magnitude of stationary response and found that within range of free roll, damping formulation has no effect on response but indicated that as angle increases, dependence of roll angle on damping moment is important[(5)].

Haddara (1989) used energy method to calculate angle dependant roll damping from analysis of free roll [(6)].

Taylon (1998) also compared analytical maximum roll amplitudes in regular waves with Wright et al. (1979) experimental data[(7)]. Taylon (2000) solved nonlinear uncoupled roll equation in frequency domain using Duffing method[(8)]. He used three widely used damping models for obtaining analytical solution.

It is clear from above studies that nonlinearity is very important to take into consideration in estimation of roll damping models for obtaining analytical solution. Those studies also indicate that damping formulation or representation is very important as estimation of maximum amplitudes vary with the type of damping presented. An effective and reliable nonlinear damping formulation is one that predicts roll response better when compared with experiments.

In present study, the effect of nonlinear damping representation on roll response in regular beam waves is evaluated. Three types of damping formulations are used for this purpose. A nonlinear ordinary differential equation with nonlinear damping and restoring moment, which represents equation of motion for roll of ship, is analytically solved using multiple scales method.

Roll response at natural roll frequency is critical one. Roll damping at natural roll frequency is estimated by analysis of free roll decay of the ship model. Linear and nonlinear damping values obtained from analysis of free roll are used to estimate maximum roll amplitude at the natural roll frequency. Analysis of free roll gives damping values corresponding to natural roll frequency. However, at design state it is difficult to obtain free roll curve due to lack of facilities to conduct experiments and/or economic constraints. In this case designer has to resort to hydrodynamic theories to estimate linear and nonlinear damping coefficients. However designer may not sure which type of damping representation has to be considered for estimation of roll characteristics of the proposed ship. It is important to verify the effect of the form of damping moment on roll response with damping moment calculated from hydrodynamic theories. Roll damping at zero forward speed consists of various components such as wave damping, skin-friction, eddy making and bilge keel damping. These components are obtained from combination of potential theory and semi empirical formulas. Wave damping as a function of wave frequency is estimated using potential theory of Ursell-Tasai method with Lewis conformal mapping. Skin friction damping, eddy-making damping and bilge keel damping are estimated using semi-empirical formulas. Maximum roll amplitude at resonance is calculated with coefficients obtained from potential theory and semi empirical formulas.

2. Mathematical model

A nonlinear ordinary differential equation with nonlinear expression for damping and restoring moment is considered to represent an uncoupled roll equation of motion of ships as follows:

$$(I_{44} + A_{44}(\omega))\theta + B(\omega)(\theta,\theta) + \Delta(K_1\theta + K_3\theta^3) = E\cos\omega_e t \quad (1)$$

where,

$E = \omega_e^2 \alpha_m I_{44}$, when θ is defined as the relative roll angle; and

$E = \Delta GM\omega_e^2 \alpha_m$, when θ is defined as the absolute roll angle.

$(I_{44} + A_{44})$ is sum of mass moment of inertia and added moment of inertia in roll, B is damping moment, which includes both linear and nonlinear damping components. Δ is displacement of ship, K_1 and K_3 are restoring moment coefficients obtained from formulation of statistical stability curve to cubic polynomial ($GZ = K_1\theta + K_3\theta^3$, where GZ is righting arm). α_m is maximum wave slope. ω_θ is natural roll frequency and ω_e is wave encounter frequency.

The following representations for damping moment are considered. They are named as B1, B2 and B3 type after Taylon[8].

$$B1(\theta) = B_{1L}(\omega)\theta + B_{1NL}(\omega)\theta|\theta|,$$

$$B2(\theta,\theta) = B_{2L}(\omega)\theta + B_{2NL}(\omega)\theta^2\theta, \quad (2)$$

$$B3(\theta) = B_{3L}(\omega)\theta + B_{3NL}(\omega)(\theta)^3.$$

If Eq. (1) with B2 type damping is divided throughout by $(I_{44} + A_{44}(\omega))$, it takes the following dimensionless form:

$$\theta + b_{2L}(\omega)\theta + b_{2NL}(\omega)\theta^2\theta + \omega_\theta^2(\omega)\theta + k_3(\omega)\theta^3 = M(t)$$

where (3)

$M(t) = K\cos(\omega_e t)$.

$K = \lambda\omega_e^2\alpha_m$, when θ is defined as the relative roll angle; and

$K = \lambda\omega_\theta^2\alpha_m$, when θ is defined as the absolute roll angle.

In the above equation,

$$\omega_\theta^2 = \frac{\Delta GM}{(I_{44} + A_{44}(\omega))}, \quad b_{2L} = \frac{B_{2L}(\omega)}{(I_{44} + A_{44}(\omega))},$$

$$b_{2NL} = \frac{B_{2NL}(\omega)}{(I_{44} + A_{44}(\omega))}, \quad k_3 = \frac{\Delta K_3}{(I_{44} + A_{44}(\omega))}, \quad \lambda = \frac{I_{44}}{(I_{44} + A_{44}(\omega))}.$$

where GM is metacenter height. λ may be considered as effective wave slope[3], more appropriately if θ in Eq.(1) is relative roll angle. Similar dimensionless equations may be also obtained with other damping representations. Where b_{2L} and b_{2NL} are dimensionless linear and nonlinear damping coefficients and k_3 is dimensionless nonlinear restoring coefficient.

3. Analytical solution of nonlinear roll equation

This section describes the analytical solution of the nonlinear roll equation with B3 type damping using the method of multiple scales method [9].

Instead of using the frequency of excitation ω_e as a parameter, introducing a detuning parameter σ , which quantitatively describes the nearness of ω_e to ω_ϕ.

$$\omega_\varepsilon = \omega_\phi + \varepsilon\sigma \qquad (4)$$

The linear undamped theory will predict unbounded oscillations when $\sigma = 0$ irrespective of how small the excitation. In the actual systems large oscillations are limited by the damping and nonlinearity. Thus to obtain a uniform valid approximate solution of this problem, excitation is ordered so that it will appear when the damping and the nonlinearity appear. This is achieved with $K = \varepsilon k$.

An approximate solution to Eq. (3) may be obtained with the method of multiple scales. Accordingly the solution may be expressed in terms of different time scales as follows:

$$\varphi(t;\varepsilon) = \phi_0(T_0, T_1) + \varepsilon\phi_1(T_0, T_1) + \ldots \qquad (5)$$

Where $T_0 = t$ and $T_1 = \varepsilon t$.

Excitation may also be expressed in terms of T_0 and T_1 as

$$M(t) = \varepsilon k \cos(\omega_\phi T_0 + \sigma T_1) . \qquad (6)$$

Substituting Eq. (5) and Eq. (6) in Eq. (3) and equating the coefficients of equal order, following equations may be obtained.

$$D_0^2\phi_0 + \omega_\phi^2\phi_0 = 0 , \qquad (7)$$

$$D_0^2\phi_1 + \omega_\phi^2\phi_1 = -2D_0D_1\phi_0 - b_{3L}D_0\phi_0 - b_{3NL}(D_0\phi_0)^3 + k_3\phi_0^3 + k\cos(\omega_\phi T_0 + \sigma T_1). \qquad (8)$$

A solution to Eq. (7) may be written as:

$$\phi_0(t) = A(T_1)\exp(i\omega_\phi T_0) + \overline{A}(T_1)\exp(-i\omega_\phi T_0).$$

Individual terms in Eq. (8) may be written as follows:

$$2\phi_0 = 2i\omega_\phi A'(T_1)\exp(i\omega_\phi T_0) + cc ,$$

$$b_L D_0\phi_0 = ib_L\omega_\phi A(T_1)\exp(i\omega_\phi T_0) + cc ,$$

$$b_{3NL}(D_0\phi_0)^3 = ib_{3NL}\omega_\phi^3 A^3(T_1)\exp(i3\omega_\phi T_0) + i3b_{3NL}\omega_\phi^3 A^2\overline{A}(T_1)A(T_1)\exp(i\omega_\phi T_0) + cc$$

$$k_3\phi_0^3 = A^3(T_1)\exp(i3\omega_\phi T_0) + 3A^2(T_1)\overline{A}(T_1)\exp(i\omega_\phi T_0) + cc,$$

where cc stands for complex conjugate of the preceding terms. In above equations D_0 and D_1 show derivation with respect to time scales T_0 and T_1 . Secular terms will be eliminated from the particular solution of Eq. (6) if A to be a solution of:

$$-2i\omega_\phi A' - ib_{3L}\omega_\phi A - i3b_{3NL}\omega_\phi^3 A^2\overline{A} + 3k_3A^2\overline{A} + \frac{1}{2}k\exp(i\sigma T_1) = 0. \qquad (9)$$

Eq. (9) is solved after writing A in the polar form as follows:

$A = \frac{1}{2}a\exp(i\beta)$, where a and β are real. After substitution of polar form 'A' in Eq. (9) and separating real and imaginary part, following equations may be obtained.

$$a' = -\frac{b_{3L}a}{2} - \frac{3}{8}b_{3NL}a^3\omega_\phi^2 + \frac{k}{2\omega_\phi}\sin(\sigma T_1 - \beta) , \qquad (10)$$

$$a\beta' = -\frac{3k_3a^3}{8\omega_\phi} - \frac{k}{2\omega_\phi}\cos(\sigma T_1 - \beta) . \qquad (11)$$

Above two equations may be transformed into autonomous systems by letting $\gamma = \sigma T_1 - \beta$.

$$a' = -\frac{b_{3L}a}{2} - \frac{3}{8}b_{3NL}a^3\omega_\phi^2 + \frac{k}{2\omega_\phi}\sin(\gamma) , \qquad (12)$$

$$a(\sigma - \gamma') = -\frac{3k_3a^3}{8\omega_\phi} - \frac{k}{2\omega_\phi}\cos(\gamma) . \qquad (13)$$

Steady state solution occurs when $a' = \gamma' = 0$, corresponds to the singular points of Eq. (10) and Eq. (11), i.e., they correspond to the solution of the following two equations:

$$\frac{b_{3L}a}{2} + \frac{3}{8}b_{3NL}a^3\omega_\phi^2 = \frac{k}{2\omega_\phi}\sin(\gamma) , \qquad (14)$$

$$a\sigma + \frac{3k_3a^3}{8\omega_\phi} = -\frac{k}{2\omega_\phi}\cos(\gamma) . \qquad (15)$$

Squaring and adding above two equations gives frequency-response relation as follows:

$$\left(\frac{b_{3L}a}{2} + \frac{3b_{3NL}a^3\omega_\phi^2}{8}\right)^2 + \left(a\sigma + \frac{3k_3a^3}{8\omega_\phi}\right)^2 = \frac{k^2}{4\omega_\phi^2} . \qquad (16)$$

Above frequency-response may be used to obtain steady roll amplitude 'a' provided numerical values of all parameters are known. σ in above equation may be considered as difference between wave encounter frequency and roll natural frequency. Above procedure may be carried out with B1 and B2 type damping to obtain corresponding frequency-response relations. Similar frequency-response relations of above type were also obtained by Talon[8] using duffing method and by Cardo.A et al. [2] using perturbation method.

4. Experiments

Measured roll responses obtained from the earlier experiments conducted at the ship maneuvering research basin of Tokyo University of Marine Science and Technology, Tokyo, Japan on a model of training ship of SHIOJI MARU has been used. The model was fitted with bilge keel. The model to ship scale ratio is 1:17.037. Principle dimensions of model and ship are shown in Table 1. The scale model with maximum length compatible with tank width was positioned transversely and its position was maintained by connecting it to the towing tank carriage pillar/post. Roll experiments were carried out with heave free arrangement. Ship model is allowed to heave with heave free arrangement. Roll data was recorded only at zero forward speed. Care has been taken so that the model will not obstruct by towing tank stationary post or fixing arrangement even the model reaches higher roll amplitudes.

Forced roll experiments in regular beam waves were conducted by varying wave frequency at constant wave height. Wave heights are varied from 30mm to 100mm with constant increment of 10mm. Natural roll period of model is about 1.5 seconds. Natural roll period is calculated from free roll curve of the model. Fig.1 shows dimensionless roll response curves. This figure clearly indicates effect of nonlinearity at resonance frequency.

Table1 Principal dimensions of model and ship

Particulars	Model	Ship
Length between p.p (m)	2.7	45.0
Breadth, molded (m)	0.587	10.0
Depth, molded (m)	0.358	6.1
Draft molded (m)	0.176	3.0
C.G above baseline (m)	0.1946	3.31
Displacement (ton)	0.150	785

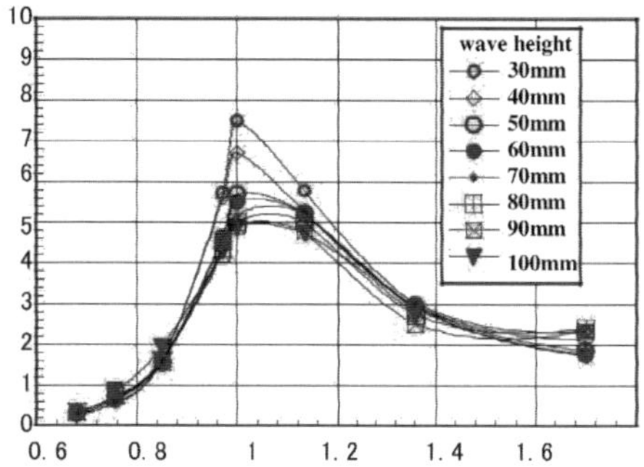

Fig.1 Dimensionless roll response curves obtained from experiments

5. Estimation of hydrodynamic coefficients

5.1 Free roll analysis

Decrement and extinction curves obtained from free roll decay analysis were used to estimate the damping coefficients. Numerical values of linear and nonlinear damping coefficients are assumed to be dependent upon the form of damping moment presented in nonlinear equation. Nonlinear damping components of B2 type and B3 type damping are obtained from formulas given by Haddara (1984)(5) for generalized damping models. Table2 shows numerical values of linear damping and nonlinear damping coefficients obtained from analysis of extinction curve with degree units.

Table 2 Numerical values of linear and nonlinear damping coefficients

Type of roll equation	Linear damping	Nonlinear damping
B1type nonlinear	0.08533	0.010986
B2 type	0.08553	0.00123
B3 type	0.08553	0.000398

5.2 Potential theory and empirical formulas

Linear and nonlinear damping coefficients obtained from analysis of the free roll are corresponds to the natural roll frequency. However numerical values of linear and nonlinear damping coefficients and effective wave slope are dependent on wave frequency. It is assumed that only wave-damping component contributes to the linear damping moment. Linear wave damping is estimated by strip method by integrating 2D hydrodynamic radiation damping coefficients over the length of the ship. Two-dimensional wave damping coefficients are calculated from Ursell-Tasai method with Lewis conformal mapping. Fig.2 shows linear damping coefficient as a function of wave frequency. It is assumed that skin friction, eddy-making and bilge keel damping components all together contribute to nonlinear part of roll damping. Skin friction damping, bilge keel damping and eddy-making damping moments are obtained from the empirical formulas of widely used Prof. Ikeda method. These components are added together to obtain total nonlinear damping moment. It may be observed that, in this case numerical value of non-linear damping coefficients is equal irrespective of the form of nonlinear damping representation. i.e., numerical value of nonlinear damping coefficient of B1, B2 and B3 type damping moments are equal. Fig.3 shows the nonlinear roll-damping coefficient as function of wave frequency.

Effective wave slope (λ) is given as ratio of mass moment of inertial and total moment of inertial (Eq.3). Added mass moment of inertial is also obtained using strip method. Ursell-Tasai method with Lewis conformal mapping is employed to obtain added mass moment of inertia. Effective wave slope is function of frequency due

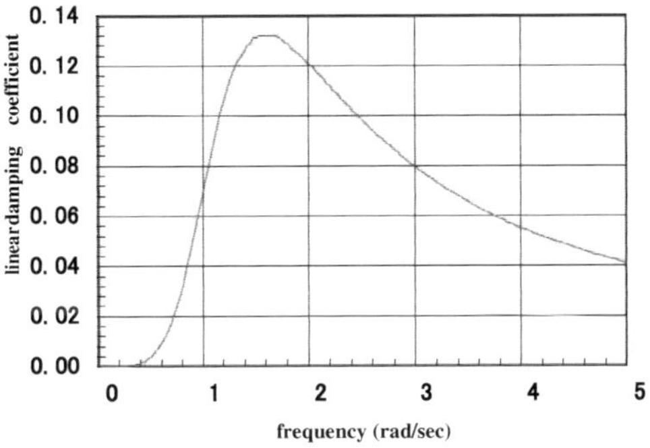

Fig.2 Linear damping coefficient vs. wave frequency

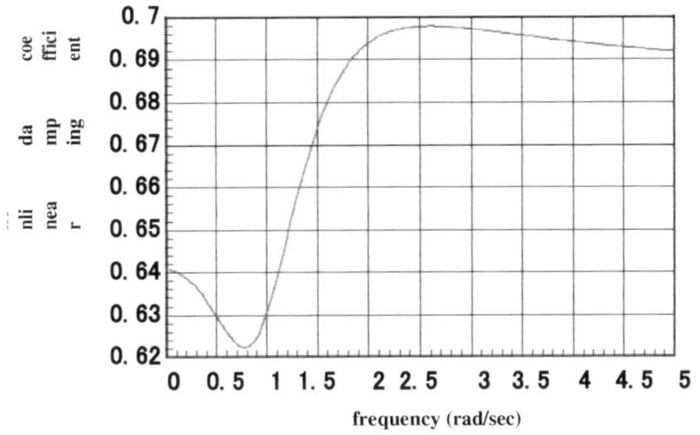

Fig.3 Nonlinear damping coefficient vs. wave frequency

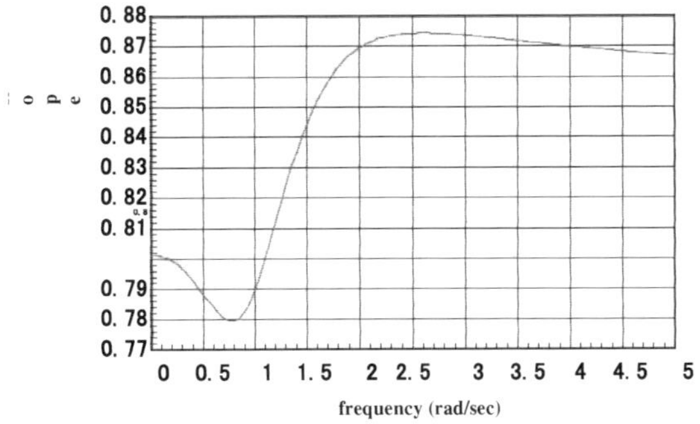

Fig.4 Effective wave slope vs. wave frequency

to frequency dependence of added mass moment of inertia. Fig.4 shows effective wave slope as a function of wave frequency. These values may be very appropriate when roll (ϕ) is defined as relative angle in Eq. (1). Values of coefficients up to 5 rad/sec are shown in Fig.2 to Fig.4. However values up to about 1.6 rad/sec are used to calculate roll amplitudes for comparing with experiments.

6. Results

Roll amplitudes are calculated using the 'frequency-response' relations obtained with the method of multiple scales; application of this method to B3 type equation was shown in Section 3. Roll amplitude with three types of damping moment are calculated. Comparison between calculations and experimentally obtained roll amplitudes at natural roll frequency are given in Section 6.1. In Section 6.1 use has been made of damping coefficients obtained from analysis of free roll decay.

Comparison of roll response curves obtained with three types of nonlinear equations and experiments are shown in Section 6.2. Response curves in Section 6.2 are calculated with hydrodynamic coefficients obtained from potential theory and semi empirical formulas. Also shown in Section 6.2 are comparison of maximum roll amplitudes between calculations and experiments.

6.1 Response at natural roll frequency (Hydrodynamic coefficients from analysis of free roll)

Linear and nonlinear damping coefficients obtained from analysis of free roll are used for this purpose. Firstly, relative roll angle is calculated analytically with three type of damping representations. For this purpose use has been made of corresponding 'frequency-response' relation (Section 3). Wave slope amplitude is added vectorically to the relative roll angle to obtain absolute roll angle. Absolute roll amplitudes are required to compare with experiment values[3]. At natural roll frequency it is assumed that, phase angle between roll and wave is zero.

Fig.5 shows the variation of maximum roll amplitude with wave height (at natural roll frequency) obtained with three types of damping representations. For comparison, experiment values are also shown on the same figure. Fig.6 shows difference between calculated maximum roll amplitudes with three damping formulations and experiment values. Negative values indicate when calculations are less than experiment values. It may be observed from these figures that, maximum roll amplitude (at resonance) calculated with B1 type damping is less than experiment values at all wave heights. Maximum deviation with B2 type damping is 15% higher than experiment. With B3 type damping, maximum deviation is around 12% higher than experiment. Difference between maximum roll amplitudes calculated either with B2 and B3 type damping is almost negligible. It may be considered that B2 and B3 type damping gives better results than B1 type damping when compared with experiment values.

6.2 Roll response curves

(Hydrodynamic coefficients from potential theory and empirical formulas)

Roll response curves are obtained using the numerical values of hydrodynamic coefficients that are shown in Figs.2 to 4. However values up to about 1.6 rad/sec are used to obtain roll response curves. Variation of restoring

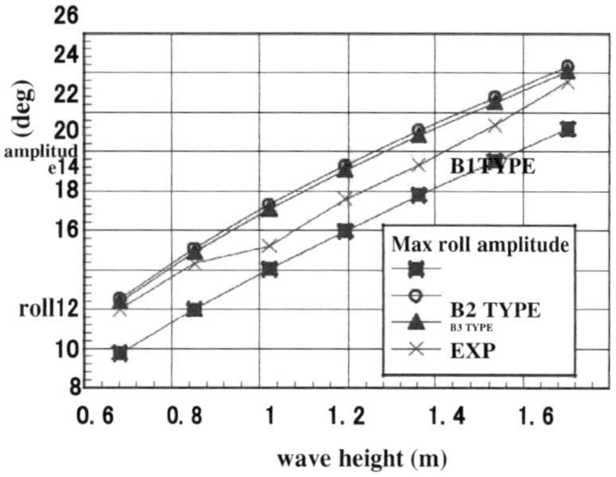

Fig.5 Maximum roll amplitudes vs. wave height

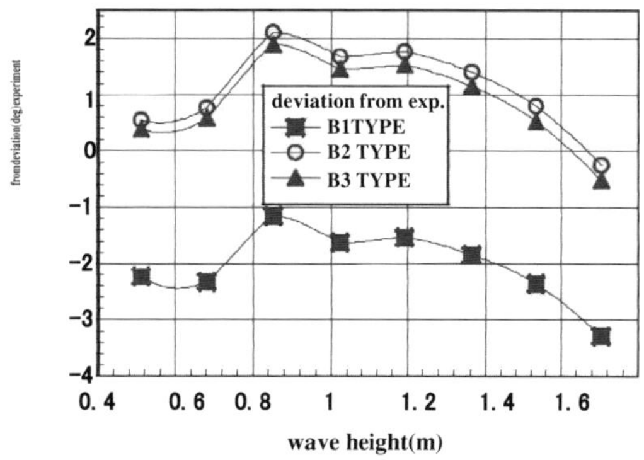

Fig.6 Difference between calculations and experiments (max roll amplitude) (negative values indicated when calculations are less than experiments)

moment coefficients (ω_φ, k_3) with frequency in nondimensionalised equation is also considered. Restoring moment coefficients (ω_φ, k_3) are defined at the Eq. (3). As phase difference between roll and wave is not know through present analytical solution, absolute values of the roll amplitude are directly calculated from the analytical solution of nonlinear roll equation with roll (φ) as absolute angle.

Figs. 7 to 9 shows roll response curves calculated with three types of damping moment with wave heights of 0.68m, 1.2m and 1.5m respectively. These wave heights are corresponding to ship scale. Also shown on these figures are roll amplitudes obtained from experiments. It may be observed from these graphs that calculated roll amplitude is same at lower and higher wave frequencies irrespective of type damping moment representation. However there is much difference between peak amplitude values at resonance frequency. Peak values (maximum roll amplitudes at natural roll frequency) are extracted from response curves and shown separately in Fig. 10. Difference between calculations and experiments are shown in Fig. 11. From these figures, it may be observed that, maximum roll amplitudes obtained with B1 type damping are less than experiment values. Maximum roll amplitudes with B3 type damping are in good agreement with the experiments. Maximum roll amplitudes obtained with B2 type damping are higher than experiments and those values calculated with B3 type damping.

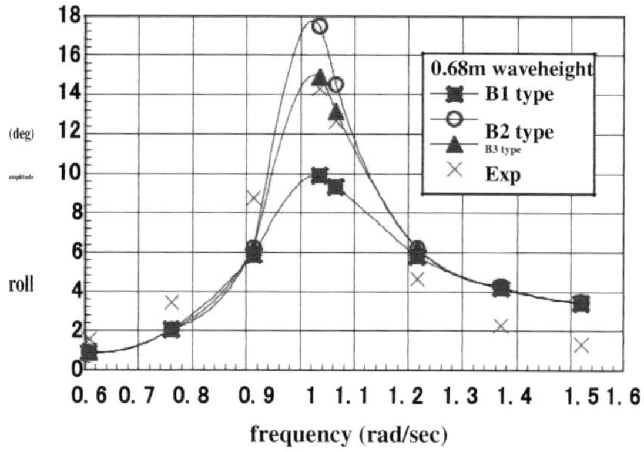

Fig.7 Roll response curves (wave height: 0.68m.)

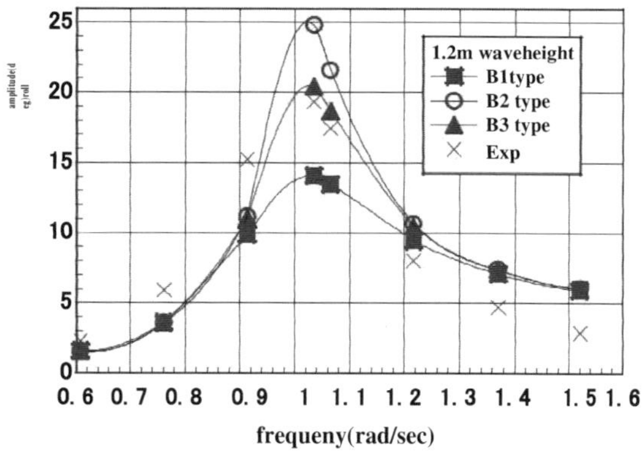

Fig.8 Roll response curves (wave height: 1.2m.)

7. Conclusions

The focus to understand ship motions and stability in the nonlinear sense is being continued. It is compelled from the compliance with ship's survivability to withstand severe weather conditions. For this purpose, a reliable nonlinear model has to be considered to assess ship's performance. In present study the effect of nonlinear damping form on roll response of a training ship is investigate using the 'frequency-response' relation from the method of multiple scales. Conclusions are made as follows:

1) Maximum roll amplitudes calculated using linear and nonlinear damping coefficients obtained from free roll analysis indicate that, linear plus velocity damping (B3 type damping) and linear plus angle dependant damping (B2 type damping) give better results than linear plus quadratic damping (B1 type damping). Difference between maximum roll amplitudes calculated either with B2 and B3 type damping is almost negligible.

2) Roll response curves are calculated by considering frequency dependence of linear damping, nonlinear damping and effective wave slope coefficients, which are obtained from hydrodynamic theories. Maximum roll amplitudes with B3 type damping are in good agreement with the experiments. Maximum roll amplitudes with B1 type damping are less than experiments where as with B2 type damping it is found higher than experiments.

It must be mentioned that the effectiveness of nonlinear model depends on form of the ship. In this study method of multiple scales along with a training ship is used for experiments. Experiments must be performed with wide variety of ship types and with higher wave heights to obtain clear understanding on the effect of damping moment form on roll response, which is reserved for further study.

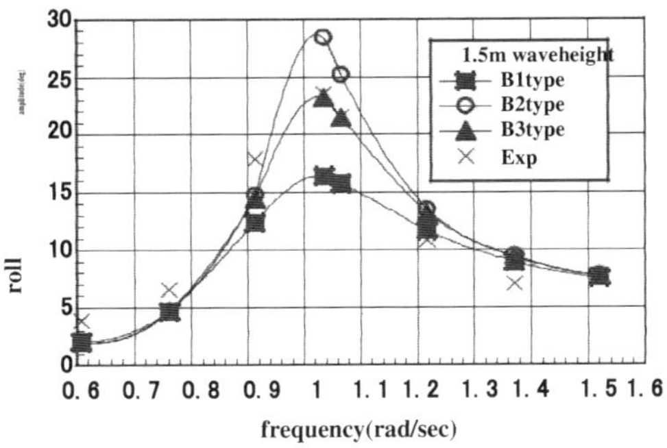

Fig.9 Roll response curves (wave height: 1.5m.)

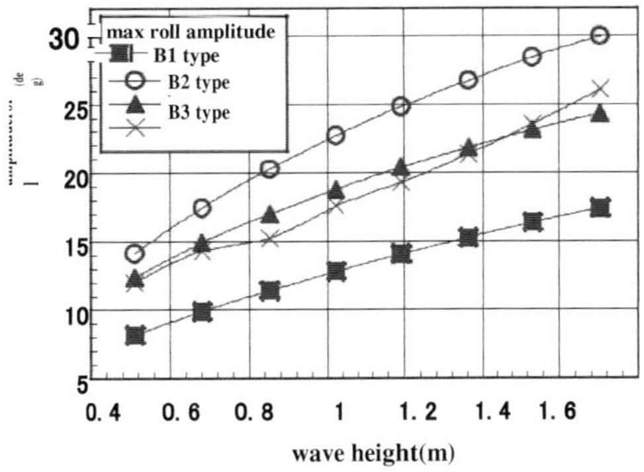

Fig.10 Maximum roll amplitudes (at natural roll frequency) vs. wave height

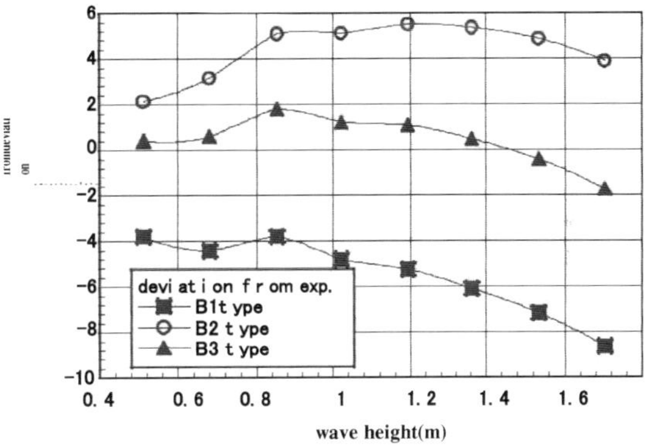

Fig.11 Difference between calculations and experiments (maximum roll amplitude) (negative values indicated when calculations are less than experiments)

8. References

1) Dalzal : A note on the form of ship rolling, Journal of Ship Research 22(3), 178-185, 1978

2) A.Cardo, M.Ceschia, A.Francescutto and R.Nabergoj : Effect of the angle-dependent damping on the rolling motion of ships in regular beam seas, International Ship Building Progress, Vol. 27, No.310, pp 135-138, 1980

3) Wright J.G. H., Marshfield W.B : Ship roll response and capsize behavior in beam seas, Trans R.I.N.A., Vol. 122, pp 129-148, 1979

4) Cardo Antonio, Francescutto Alberto : On damping models in free and forced rolling motion, Ocean Engng. Vol.9, No.2, pp 171-179, 1882

5) Haddara M.R : A note on the effect of damping moment form on rolling response, International Ship Building Progress, Vol.31, No.363, pp 285-290, 1984

6) Haddara M.R : A study of the angle dependence of roll damping motion. Ocean Engng. Vol.16, No.4, pp 411-427, 1989

7) Taylon : Analysis and validation of ad hoc nonlinear roll motion models, ARI-An International Journal for Physical and Engineering Sciences, Vol. 51, pp 113-119, 1998

8) Taylon : The effect of nonlinear damping and restoring in ship rolling, Ocean Engng.Vol. 27, pp 921-932,2000

9) A.H. Nayfeh, T.Mook : Nonlinear oscillations, John Willey&Sons,Edition1995

9. AUTHOR BIOGRAPHY

S.Rao Tata is employed with Lloyds Register Asia, Singapore in the South Asia Technical support Office. He is responsible for plan approval tasks in the statutory computational matters. His previous experience includes in the involvement of marine consultancy work, research and development in marine hydrodynamics.

The Royal Institution of Naval Architects
8-9 Northumberland Street
London, WC2N 5DA
United Kingdom

ISBN 978-1-5108-8307-9

Design & Operation of Wind Farm Support Vessels 2019

London, United Kingdom
27 February 2019

The Royal Institution of
Naval Architects